Rick **Armstrong** | Elani **McDonald** | Jennifer **Barnett** | Rac

Science 3
for the international student

PROPERTY OF KING FAHAD ACADEMY

Australia • Brazil • Japan • Korea • Mexico • Singapore • Spain • United Kingdom • United States

Science 3 for the International Student
2nd Edition
Rick Armstrong
Rachel Whan
Elani McDonald
Gareth Jones
Jenny Sharwood
Neil Champion
Julie Hall
Tanith James
Katrina Mitchell

Publishing editor: Sarah Craig
Editor: Catherine Greenwood
Proofreader: Stephanie Ayres
Cover design: Aisling Gallagher
Text design: Aisling Gallagher
Cover image: Sean Locke/Stocksy United
Permissions researcher: Sian Bradfield
Production controller: Emma Roberts
Typeset by: Macmillan Publishing Solutions

Any URLs contained in this publication were checked for currency during the production process. Note, however, that the publisher cannot vouch for the ongoing currency of URLs.

© 2016 Cengage Learning Australia Pty Limited

Copyright Notice
This Work is copyright. No part of this Work may be reproduced, stored in a retrieval system, or transmitted in any form or by any means without prior written permission of the Publisher. Except as permitted under the *Copyright Act 1968*, for example any fair dealing for the purposes of private study, research, criticism or review, subject to certain limitations. These limitations include: Restricting the copying to a maximum of one chapter or 10% of this book, whichever is greater; providing an appropriate notice and warning with the copies of the Work disseminated; taking all reasonable steps to limit access to these copies to people authorised to receive these copies; ensuring you hold the appropriate Licences issued by the Copyright Agency Limited ("CAL"), supply a remuneration notice to CAL and pay any required fees. For details of CAL licences and remuneration notices please contact CAL at Level 15, 233 Castlereagh Street, Sydney NSW 2000, Tel: (02) 9394 7600, Fax: (02) 9394 7601
Email: info@copyright.com.au
Website: www.copyright.com.au

For product information and technology assistance,
in Australia call **1300 790 853**;
in New Zealand call **0800 449 725**

For permission to use material from this text or product, please email
aust.permissions@cengage.com

National Library of Australia Cataloguing-in-Publication Data
Armstrong, Rick, author, editor.
Science 3 for the international student / Rick Armstrong, Rachel Whan, Elani McDonald, Gareth Jones, Jenny Sharwood, Neil Champion, Julie Hall, Tanith James, Katrina Mitchell.
    2nd edition.
    9780170353540 (paperback)
    Science for the international student.
    Includes index.
    Series editor: Rick Armstrong.
    For secondary school age.
Science--Textbooks.
Science--Study and teaching (Secondary)--Australia.
International baccalaureate--Study guides.

Whan Rachel, author. McDonald, Elani, author.
Jones, Gareth, author. Sharwood, Jenny, author.
Champion, Neil (Neil Douglas), author. Hall, Julie, author.
James, Tanith, author. Mitchell, Katrina, author.

500

Cengage Learning Australia
Level 7, 80 Dorcas Street
South Melbourne, Victoria Australia 3205

Cengage Learning New Zealand
Unit 4B Rosedale Office Park
331 Rosedale Road, Albany, North Shore 0632, NZ

For learning solutions, visit **cengage.com.au**

Printed in China by 1010 Printing International Limited.
1 2 3 4 5 6 7 20 19 18 17 16

Contents

About the authors vi
How to use this series vii
Introduction ix

UNIT 1 INDIGENOUS KNOWLEDGE   1

Introduction 2
Indigenous knowledge 2
Are Indigenous knowledge and modern scientific knowledge compatible? 4
Indigenous knowledge as holistic knowledge 6
We see the world through our cultural values 6
Oral transmission of knowledge and use of mythical stories 8
Indigenous peoples' sourcing of food 9
Indigenous knowledge and diet 11
Cooking with heated stones 13
      Investigation 1.1: Cooking with hot stones 14
Indigenous knowledge about medicines 15
      Investigation 1.2: Properties of chamomile 16
The health of Indigenous people 17
Indigenous peoples and intellectual rights 18
Indigenous knowledge about navigation 20
The Kayapo: A modern case study 20
Impact of modern conservation approaches on Indigenous people 22
Unit questions 24

UNIT 2 ADAPTATIONS OF ORGANISMS   25

Introduction 26
Adaptations 26
Adaptations in bears and marsupials 29
Evolution by natural selection 32
Artificial selection 36
Plant adaptations to dry environments 37
      Experiment 2.1: Observing stomata 38
      Investigation 2.1: Controlling water loss from leaves 38
Human impact on natural communities 40
Unit questions 42

UNIT 3 LOOKING AFTER OURSELVES   43

Introduction 44
Adolescents' health – the global challenge 44
You as an adolescent 45

Physical health 48
Diet for adolescents 50
Looking after the emotional and social you 52
The brain and adolescence 54
Drug misuse and abuse.. 55
          Investigation 3.1: What factors affect lung capacity? 62
Unit questions 63

UNIT 4 USING METALS   65

Introduction 66
Metals 66
Types of metals 69
          Experiment 4.1: Colours of transition metal compounds 72
Alloys 73
Non-metals and metalloids 73
How do metals react? 75
          Investigation 4.1: Investigating rust 77
          Experiment 4.2: A potato battery 79
          Investigation 4.2: Making electricity 80
          Experiment 4.3: Electroplating 81
Unit questions 84

UNIT 5 USEFUL CHEMICAL REACTIONS   87

Introduction 88
Chemical and physical change 88
          Experiment 5.1: Observing chemical reactions 89
Gases produced in chemical reactions.. 91
Representing chemical reactions 93
Types of chemical reactions 95
          Experiment 5.2: Heating copper carbonate 97
          Experiment 5.3: A metal displacement reaction 99
          Investigation 5.1: Predicting reactions 100
More specific reactions...101
          Experiment 5.4: Neutralisation reaction 105
Unit questions 107

UNIT 6 WAVES: LIGHT AND SOUND   109

Introduction 110
What are waves? 111
Sound waves 112
          Experiment 6.1: Teacher demonstration – sound..112
          Experiment 6.2: Using an oscilloscope to show volume of sound waves 114

Investigation 6.1: Pitching it right 116
Investigation 6.2: Sound travelling through materials 117
Experiment 6.3: The speed of sound... 119
Light..120
Experiment 6.4: Mirror image... 121
Experiment 6.5: How many images?... 122
The electromagnetic spectrum 125
Modern communication technology 128
Unit questions 130

UNIT 7 ASSISTIVE TECHNOLOGY 131

Introduction 132
Understanding disability 132
Disabilities from wars 136
Help from simple machines 137
Investigation 7.1: The functioning of bicycle gears................................. 144
Electrical control systems 145
Unit questions 150

UNIT 8 THE SUN AND MOON IN OUR LIVES 151

Introduction 152
Influence of the Sun on Earth..152
Experiment 8.1: Measuring the diameter of the Sun using a pinhole camera...........154
Investigation 8.1: The size of the image in a pinhole camera 156
Influence of the Moon 159
Phases of the Moon 164
Eclipses 166
Tides..169
Unit questions 171

Appendix 1: Approaches to Learning (ATL) framework in MYP Sciences........................173
Appendix 2: MYP Science 3 assessment criteria...175
Appendix 3: Guidance on carrying out and writing up MYP 3 scientific investigations (criteria B and C) 177
Appendix 4: Articulating the conceptual framework in MYP Sciences........................179
Glossary 183
Index 188

About the authors

Authors

Rick Armstrong (series editor)
Rick Armstrong has been involved with MYP sciences guide writing since 1994. He has led science workshops in all International Baccalaureate regions, as well as Approaches to Learning workshops. He has been involved with moderation, school visits and authorisations, and has experience as a DP examiner. Rick is currently a freelance educational consultant in Madrid, Spain.

Jennifer Barnett
Jennifer Barnett has been involved with the MYP since 2005 and is a sciences workshop leader and school authorisation team member. Recently, Jennifer was chosen to be part of the International Baccalaureate service 'Building Quality Curriculum' to evaluate teachers' unit plans for school authorisation. She has also led a number of local and state workshops on incorporating technology in the science classroom and differentiating science for exceptional students. She currently teaches integrated sciences to MYP years 1–3 in Austin, Texas.

Elani McDonald
Elani McDonald has dedicated her entire teaching career to working in IB schools. She is a workshop leader and an MYP visiting team member. She has been involved in monitoring and moderation of assessment and was involved in writing the Science and Personal Project 2014 guides, as well as the 2014 teacher support material for physics. Elani is dedicated to making learning relevant and engaging and was shortlisted for the TES maths teacher of the year award in 2014/15. Elani is teaching mathematics and sciences full-time and doing consultancy work part-time.

Rachel Whan
Rachel Whan studied veterinary science before completing a Graduate Diploma in Education. Since then she has taught chemistry, biology, human biology, science, mathematics and agriculture in a variety of schools in Queensland and Western Australia. Rachel has taught in MYP schools since 2006 and has been the Science Team Leader at St Brigid's College in Lesmurdie, Western Australia since 2007.

How to use this series

The *Science for the international student* series provides students with a variety of engaging and stimulating formats for learning, understanding and immersion in both the Middle Years Programme (MYP) philosophy of the International Baccalaureate (IB) and the science content. The features of the student book have been specifically designed to support this and to deliver exciting content in a variety of ways.

Specific MYP features

Each unit begins with a unit opening page that specifies:
- the key concept that is covered in the unit
- the related concepts that are covered in the unit
- the Global Context of the unit
- the Statement of Inquiry
- inquiry questions, divided into factual, conceptual and debatable questions.

Key and related concepts

Each unit is based around one *key concept* of an enduring transdisciplinary nature and a small number of *related concepts* designed to help frame the unit in the minds of the students.

Global Context

Students will be encouraged to see science in the *global context* of its ability to provide a basis for creative inventions that are capable of enriching our lives in areas such as space, materials, sports and medicines.

Statement of Inquiry

The *Statement of Inquiry* drives the unit and is strongly related to the units' concepts and context.

The inquiry questions are divided into factual, conceptual and debatable questions. Factual questions are related to the unit content, conceptual questions are related to the unit concepts and debatable questions are related to both and designed to stimulate deeper thinking.

Performance assessment tasks

Opportunities for assessment tasks occur throughout each unit and these are each identified by a *performance assessment task* icon.

The *summative performance assessment task* associated with the Statement of Inquiry is identified at the beginning of each unit. The criteria assessed by the assessment task are also identified.

Approaches to Learning

Opportunities to develop and apply *Approaches to Learning* skills are identified by an 'ATL' icon. Teachers can use these prompts to discuss and reinforce learning strategies.

Investigation

Investigations challenge students to design and perform their own experiments either individually or in groups. Investigations are designed to satisfy criteria B and C.

Experiments

Experiments provide students with the opportunity to develop and practise their skills by following processes and procedures, to discover information for themselves and to build a greater understanding of, and interest in, scientific concepts. Experiments are designed to satisfy criterion C.

Taking action

Taking action suggestions are identified by a 'TA' icon and are designed to satisfy the MYP requirements for service as action.

Other features

Review

Review boxes contain questions and break the content into smaller sections, allowing students to review what they have learnt so far.

Activity

Activity boxes reinforce or develop concepts and skills through short, fun and hands-on activities.

Weblinks

Weblinks are identified by an icon and direct students to exciting websites to further explore the world of science.

Unit questions

Unit questions conclude each unit. They include review questions sorted under the MYP assessment criterion A, levels 1–8. Reflection questions are included to review the concepts underpinning the unit, to encourage further consideration of the debatable inquiry questions, and at times to consider further lines of inquiry.

NelsonNetBook

The *Science for the international student* NelsonNetBook is an interactive ebook that can be used online or offline. It is compatible with interactive whiteboards, computers and tablets, with optional Web 2.0 functionality for class groups. Students can add highlights, annotations, audio and video clips, and weblinks, and teachers can use it to share their personalised version with the class.

Visit the NelsonNet portal at www.nelsonnet.com.au to find out more, register, or log in if already registered.

NelsonNet teacher website

The NelsonNet teacher website contains further valuable advice, including draft MYP unit plans covering the first two pages of the revised MYP planner, and also a curriculum overview as required by the IB. Other resources include blackline masters (BLMs) containing possible further experimental work and classroom activities, ideas for further resources, and further advice relating to teaching in a conceptual way and for the use of the Approaches to Learning framework. Answers are also provided for all questions, as well as a list of extra resources for each unit.

Contact your sales representative for information about access codes.

Introduction

To the student

We hope you will enjoy using this exciting student book, which has been designed to provide an up-to-date science experience around the principles of the new enriched Middle Years Programme (MYP) offered by the International Baccalaureate (IB). You are likely to already be an experienced MYP student, proud of being an *internationally minded* student, and familiar with the distinctive way MYP students work in science. These revised books provide a greater emphasis on the global contexts for learning in science, ranging from the challenge to provide better and more equal access to medicines worldwide, to considering global environmental challenges such as global warming. The books emphasise investigative and experimental work and expect you to work and think like a real scientist. As you will be well aware, the MYP is also about encouraging you to develop effective learning skills that will stay with you for life, and you will see in these books many suggestions to help you with this challenge. We wish you all the success possible with MYP Science and beyond.

To the teacher

We have reviewed our original series, published in 2010–2011, to take account of the innovative developments and improvements in the MYP. In this new edition, we have deepened our coverage of MYP principles within each unit. The units are now much more contextual and more explicitly driven by the Statement of Inquiry. As you will be aware, the IB has attempted to give schools more flexibility in their delivery of the MYP and there certainly is no 'correct' model of how to put the MYP into practice. For that reason, we feel we should explain some of our approaches to constructing our units.

1. **Conceptual framework:** We have closely followed the suggested framework but have added a small number of extra related concepts that will be useful to teachers and will allow coverage of the US cross-cutting concepts. We have also used concepts from other subjects when we felt their use would enhance the unit. Importantly, we accept that the key to teaching conceptually lies in appropriate classroom practices. To help this practice, we have included activities and questions to help strengthen students' understanding of the conceptual framework as well as some further guidance in the teacher materials.
2. **Content:** We have included academically challenging content that will provide an effective transition from MYP 1 through to 5, to the new e-examinations, and to higher study in the Diploma Programme (DP) or in other national systems. This content should also help teachers meet the requirements of local curricula. We have covered all the expected content for MYP Sciences e-examination in Books 4/5. Some of this content is also covered in more detail in Books 1, 2 and 3. We have ensured that the scope and sequence of our MYP Books 1–5 is well thought out and offers a coherent framework for the development of deep understanding based on the big unifying concepts in science.
3. **Global Contexts:** The development of the Areas of Interaction into the Global Contexts is very liberating and opens the door for much more creative uses of contexts in the planning of MYP units. To take advantage of this potential, we have associated the Global Context chosen for the unit with a more specific 'exploration into' statement. This 'exploration into' feeds clearly into the Statement of Inquiry for each unit. This has helped us to make the science content up to date, interesting and relevant to the real world.

4   **Statements of Inquiry:** We have written simple and clear Statements of Inquiry that are understandable to students and to teachers. We have been flexible in relation to trying to build all the chosen concepts into the Statements of Inquiry. Our priority has been to ensure that the Statement of Inquiry is easy to understand, has a conceptual feel, and, importantly, relates to the chosen Global Context.

5   **Assessment tasks:** Most science units will require more than one summative performance assessment task because it is artificial to try to bring together a number of the sciences criteria in one task. Therefore, most units include assessments relating to investigation work (criteria B and C), a performance-type task relating to the impacts of science (criterion D) and end-of-unit questions to assess knowledge and understanding (criterion A). At the beginning of each unit, you will see a summative performance assessment task that relates closely to the Statement of Inquiry. We have given this task the most authentic performance nature possible. Other performance assessment tasks are included in each unit that can be used summatively or formatively. We expect that not all of the assessment suggestions will be used for summative purposes.

6   **Approaches to Learning:** We are very impressed by the revised Approaches to Learning framework based on the ten clusters of ATL skills. We understand that the effective implementation of ATL is a whole-school challenge but have made suggestions for when teachers can explicitly introduce these skills and dispositions, both as part of summative assessment tasks, and also more generally in their daily teaching. You will also see a simplified ATL framework in the appendices that we think will be of great help to teachers.

7   **Service learning:** We have also suggested a possible service learning activity (labelled 'TA' (Taking action)) for each unit.

The NelsonNet teacher website contains draft MYP unit plans, curriculum overviews, BLMs for experimental work and classroom activities, ideas for interdisciplinary tasks, further resources and advice for using the ATL framework, and answers to all questions.

We realise there may seem to be an inherent conflict between the idea of teachers working in a creative and collaborative way to produce MYP units of work and the use of a textbook. Schools will use this book in different ways. Some new schools might find it an invaluable stepping stone to getting an MYP Sciences programme up and running. Others may use it to enhance their existing courses. We encourage you not to use these books the way traditional textbooks have been used. Be creative, add to them, choose the bits you like, encourage the students to interact with them. They are there to help students in their deep learning of science, to encourage their interest and motivation. We hope the availability of materials of this kind will make your life as the teacher a little easier and give you more time to focus on the actual teaching and learning. Enjoy them.

Rick Armstrong (Series editor)

UNIT 1

INDIGENOUS KNOWLEDGE

KEY CONCEPT
Cultures

RELATED CONCEPTS
Environment

Balance

Perspective

GLOBAL CONTEXT
Personal and cultural expression – an exploration into Indigenous peoples' knowledge systems

STATEMENT OF INQUIRY
Indigenous peoples possess detailed and invaluable knowledge that allows them to live in a sustainable way with their local environments.

INQUIRY QUESTIONS

FACTUAL
1. What knowledge do Indigenous people have in relation to food, medicine and navigation?
2. What are the names of some Indigenous people who live in your region?

CONCEPTUAL
3. How does Indigenous knowledge compare with modern scientific knowledge?
4. What are the main challenges facing Indigenous people in the 21st century?

DEBATABLE
5. Is it possible for Indigenous peoples to live in the modern world while preserving traditional values and customs, sometimes called 'two-eyed thinking'?

Introduction

Indigenous peoples are people who have lived in a region for a very long time, and who have preserved their distinctive local cultures and traditions. Many Indigenous peoples live in a way that is very close to their traditions and has changed very little over the past hundreds or thousands of years. Other Indigenous peoples mix traditional and modern practices in their lives.

Indigenous peoples make up a significant percentage of the world's population. It is estimated that there are about 5000 different tribes of Indigenous peoples in the world: about 350 million people in total. The knowledge they have, particularly about local **ecology**, is immense and complex. In recent years, we have started to appreciate the value of this knowledge more and more.

Scientific knowledge of Indigenous peoples

Research
Choose one group of Indigenous people and carry out research into their customs. As a class, you might decide to concentrate on one particular group of Indigenous people, possibly a group that lives in your country. Concentrate on a small number of aspects of their lifestyle, such as their food, family structures, agriculture, hunting, clothing, use of herbal medicines and cooking. Also research the issues this group of Indigenous people experience from being part of modern society.

Your task
Write an article for a newspaper, or prepare a TV documentary, on the customs of this group of Indigenous people. Include a discussion on why it is important to understand that the implementation of modern scientific and technological developments does not always work well in Indigenous cultures.

Go to http://mypsci3.nelsonnet.com.au and click on **Survival international** to see a list of some of the Indigenous peoples facing the most significant challenges.

CRITICAL THINKING
Recognising our personal cultural assumptions and biases

Indigenous knowledge

Reflection about learning

ACTIVITY

THINK–PAIR–SHARE
Brainstorm about Indigenous peoples you have some knowledge of. Where do they live? What are their customs? What beliefs do they have? How well have they been treated by newer peoples (often Europeans) who have taken over their lands? What problems are they facing in the modern world? What contributions do they make?

- Work in pairs to share and discuss your findings.
- Pair up with another group and share and discuss your findings.

Indigenous peoples have accumulated an enormously valuable and complex knowledge of their local environments. They have expert knowledge of local materials, weather, plants, animals and rivers/waterways. They often have advanced understanding of astronomy and navigation techniques. They have developed distinctive local tools, forms of food preparation, shelter, medicines and language. Traditionally this knowledge is not written down. It is passed from generation to generation through hands-on experience, art, storytelling, dance and music.

FIGURE 1.1 Indigenous customs are important for (a) medicines, (b) transport and (c) food.

FIGURE 1.2 Dancing is a powerful way of communicating information.

In the past, there has been a tendency to think of Indigenous cultures as primitive. As a consequence, many Indigenous cultures suffered when colonising people took over their lands. This attitude is now changing and the value of Indigenous knowledge is increasingly respected. In particular, we value the ability of Indigenous peoples to live in a **sustainable** way in their local environment – a knowledge we seem to have lost in modern times. On the other hand, we must be careful not to over-romanticise all aspects of Indigenous cultures. Indigenous people have had to struggle against climate extremes, changing local ecosystems, limited diets, poor health, and inter-tribal conflicts.

Are Indigenous knowledge and modern scientific knowledge compatible?

The origin of modern science is debatable. Some people see the origins of modern science in the ancient Sumerian civilisation of 4000–2000 BCE and their ideas on astronomy and mathematics. Others see it beginning with the Greeks and their ideas on the atom. Many people link it to the scientific revolution in Europe over the 16th–18th centuries, which was associated with more sophisticated ideas on astronomy, gravity, theories of what matter is made of, the human body and electricity. Other people trace the modern scientific method back to Ibn al-Haythan in the Golden Age of Arabic science (Figure 1.3a).

In *Science 1 for the international student* Unit 1, you studied the use of the scientific method and how this way of working is reflected in MYP Sciences criteria B and C.

The scientific method involves a specific research question that has arisen from previous observations or considerations. It builds upon previous knowledge and usually involves the formulation and testing of a hypothesis. It involves the design of experiments with control of variables (fair testing) and very careful consideration of the evidence. An important aspect of the scientific method is that results are published and other scientists should be able to reproduce the method and the results.

FIGURE 1.3 (a) Ibn al-Haythan and (b) Isaac Newton were influential in the development of the modern scientific approach.

The way in which Indigenous peoples learn about their environment is likely to be different from the approach of modern science. The development of Indigenous knowledge doesn't insist on a theoretical explanation. It is not published in scientific journals. It is not normally the result of controlled experiments.

But there are similarities with modern science — Indigenous knowledge has developed from careful observation, questioning, trial and error, prediction, problem solving, interpretation and adaptation over a period of time. Indigenous knowledge is likely to have built up over hundreds, in some cases thousands, of years. Indigenous knowledge is often very practical and is usually locally based.

Both modern science and Indigenous knowledge are valid ways of making sense of the world around us. They often complement each other, particularly in areas such as health care. If you study the IB Diploma Programme you will study Theory of Knowledge (TOK). In TOK, you will consider the different ways we develop our understanding of the world.

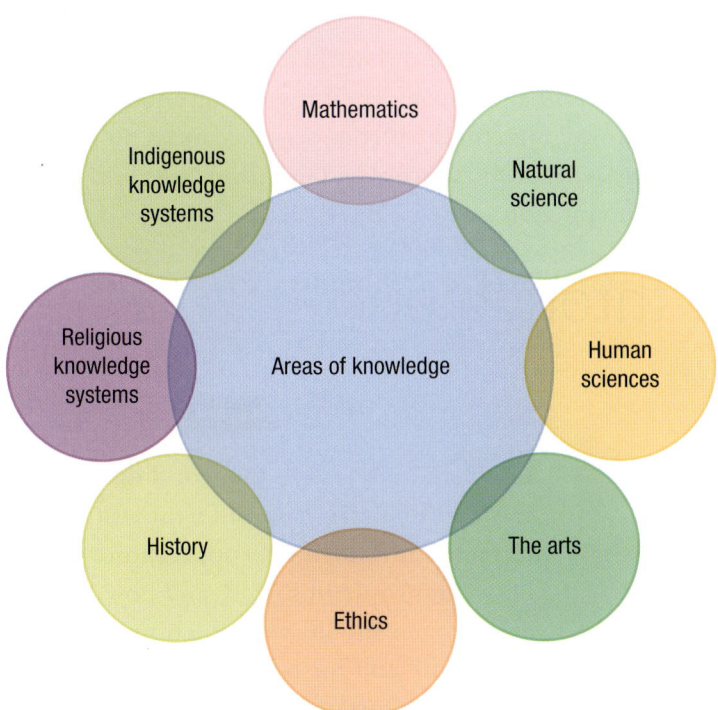

FIGURE 1.4 Areas of knowledge studied in the Diploma Programme

FIGURE 1.5 A Shuar Indian woman in Ecuador harvesting the medicinal apaye plant to treat amoeba infections

ACTIVITY: Further comparison of Indigenous knowledge with scientific knowledge

In 2008, UNESCO published a set of seven posters on the theme of Indigenous knowledge for use in education. Go to their website and choose one of the posters. Read through the poster in small groups and prepare a 3-minute presentation to your class summarising the message of the poster.

Go to http://mypsci3.nelsonnet.com.au and click on **UNESCO posters** to view the UNESCO posters.

Indigenous knowledge as holistic knowledge

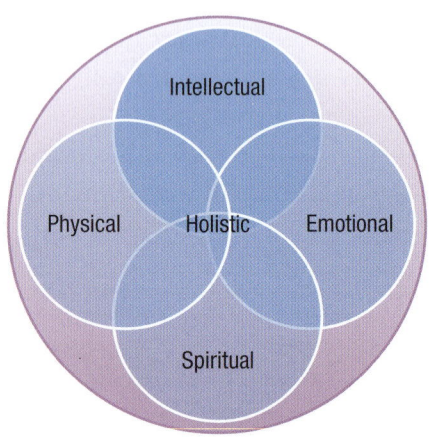

FIGURE 1.6 A holistic view

The knowledge in an Indigenous culture tries to link everything together. For example, Indigenous peoples tend to see the individual, family, society and nature as a related whole. We call this **holistic** knowledge. For example, Indigenous medicine treats the whole person, including their psychological, emotional, social, spiritual and nutritional needs. It does not just give a medicine to take away the symptoms. Interestingly, in recent times many people are turning to holistic medicine, which has a similar philosophy.

A holistic view of the world leads to an enormous respect for nature. Compare this with our modern world where we sometimes do considerable damage to local ecosystems, such as forests and grasslands, for short-term economic gain.

We see the world through our cultural values

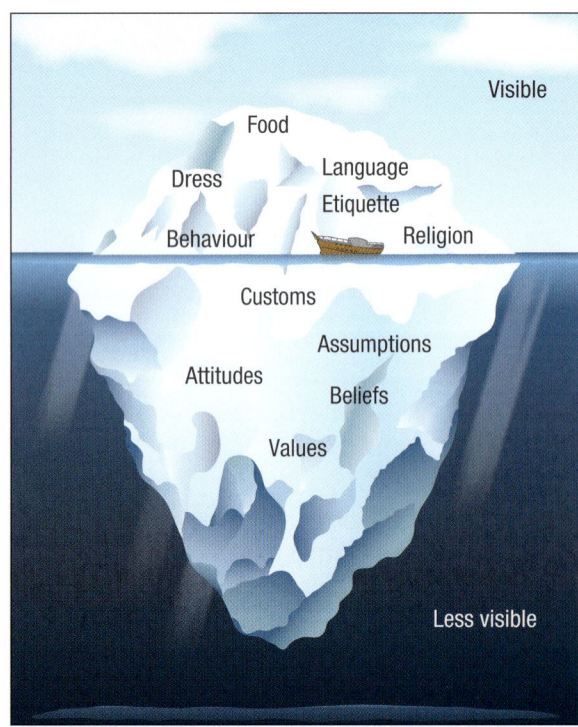

FIGURE 1.7 The culture iceberg

Different cultures see the world differently. Each culture may have different attitudes to family life, learning, cooperation between people in society, and how men and women should behave. This can make it very difficult for people of one culture to fully understand the customs and knowledge of those from another culture. We are all trapped to some extent by our cultural perspectives. The metaphor of culture being like an iceberg is often used to show why people find it difficult to understand other cultures.

Many students in international schools have experienced moving from one country to another, from one group of classmates to another. These students will probably have experienced the sensation of everything feeling different and strange. Perhaps they have had a feeling of not understanding the new culture or not feeling understood by people in the new culture.

FIGURE 1.8 Students from different cultures

ACTIVITY: Comparing cultures

In small groups, consider peoples from other cultures you know, or other cultures you have lived in. Can you think of ways in which they see the world or behave differently? Do they have different values in relation to family life, respect for elders, learning, relationships between males and females etc.? What different customs do they have? How did you feel when (or how do you think you might feel if) you entered another culture?

- Make a table of the differences you have noticed.
- Make another table of the similarities you might see.
- Describe how you felt when (or how you might feel if) you entered a different cultural world.
- What advice do you have for someone going to live or work in a new culture?

REVIEW

1. How would you define the term 'Indigenous peoples'? Give four examples of Indigenous peoples from around the world.
2. Describe ways in which Indigenous peoples have developed their knowledge.
3. Discuss ways in which modern science and Indigenous knowledge are:
    a similar
    b different.
    Construct a table to show your answer.
4. What does the term 'holistic' mean? Why do we say that Indigenous knowledge is 'holistic'?
5. We say that different cultures have 'different world views'. What does this mean? Can you think of some examples?

Oral transmission of knowledge and use of mythical stories

Knowledge in Indigenous cultures is transferred in very special ways, usually by elders. The information is carefully passed down from generation to generation. This knowledge often has a spiritual dimension, something that people from other cultures often find difficult to relate to. Mythical stories are often used to help explain important ideas. What stories and information have been passed on to you from older generations? In a recent initiative to improve the health of Inuit people in Pangnirtung, Canada, the emphasis was placed on involving the elder people in storytelling about traditional diets.

Creation stories and Mother Earth

Some of the stories that Indigenous peoples tell are about the creation time – the time when the land and the landscape, and its animals, plants and people came into being (Figure 1.9). They also tell the people about their connection to the land and how they must look after it and each other.

FIGURE 1.9 A panel from the Aztec creation myth

ACTIVITY: How was the Earth formed?

What is your understanding about how the Earth and its inhabitants were formed? What other explanations are you familiar with? In pairs, choose a creation story from an Indigenous culture and create a poster that explains what happened in the creation story when the Earth and its inhabitants were created.

To Indigenous peoples, the land is the source of their life and it is where they return to when they die, so it has to be treated with respect. Many Indigenous peoples worldwide use the phrase 'Mother Earth' to describe this relationship. As part of this close relationship, they have a detailed knowledge of the plants and animals, the seasons and the landscape and how these affect them. It is the responsibility of each person to look after the land. The idea of individuals owning land can be strange for Indigenous peoples.

To read a useful summary of creation stories, go to http://mypsci3.nelsonnet.com.au and click on **Creation stories**.

Indigenous peoples' sourcing of food

> Food will be what brings people together.
>
> Jones Ignace, Elder of the Secwepemcs, an Indigenous people of British Columbia, Canada

Hunter-gatherers

Indigenous food systems depend on the local environment. They rely on detailed knowledge of what will grow well or what is available for hunting and fishing in particular seasons, what is sustainable, and what is healthy. Early humans were hunter-gatherers and often lived a nomadic lifestyle, such as that of the Kalahari Bushmen of southern Africa (Figure 1.10), who feed themselves by hunting wild animals and gathering wild plants.

Gathering and growing

About 10 000 years ago, some peoples learnt to domesticate wild animals and plants. They produced their own food. They became farmers. This led to permanent settlements of people and eventually to more sophisticated technology.

FIGURE 1.10 The Kalahari Bushmen are hunter-gatherers.

Present-day farming methods and food production are the result of generations of investigation and experimentation. Domesticated corn (maize), beans and squash were developed in South America and Mexico. Their use spread north into North America, then later to Europe.

These three crops have been referred to by Indigenous North American peoples as the 'three sisters' because the three are interdependent (Figure 1.11). The corn stalks provide support for the beans. The beans are legumes and contain nitrogen-fixing bacteria in their roots, which provide valuable nitrogen to the soil. Squash, which is planted between the rows of corn, reduces water loss and keeps weeds out. This is called **companion planting**.

Today, farmers and gardeners make use of the concept of companion planting. For example, marigolds are often planted with vegetable patches to repel whiteflies and beetles, as well as some weeds. There are many other instances of companion planting.

FIGURE 1.11 Growing the 'three sisters' together: corn, beans and squash are interdependent.

Rice and fish have been grown together in China for over 1000 years (Figure 1.12). The fish eat the rice pests, and the rice provides a good environment for the fish. This reduces the need for pesticides by around 70% and the need for chemical fertiliser by over 20%.

Indigenous farmers in Thailand have developed very complex systems of landscaping to include gardens for their homes, land for grazing cattle, rice paddies, and trees to provide a path for animals. Hence, the community meets its needs without causing destruction or affecting local wildlife.

Development of hunting techniques

Elaborate technologies for hunting have been developed. Australian Aboriginal people use the boomerang (Figure 1.13). The boomerang is a wooden implement with a number of uses, particularly for hunting and fighting. Boomerangs are probably best known because they can return to the thrower. Returning boomerangs are shaped like two wings joined together at an angle. In cross-section, they have the same aerofoil shape as the wing of a plane. One end is twisted slightly upwards; the other end is twisted slightly downwards, causing it to spin. It is this spinning action that causes the boomerang to return to the thrower.

Other Indigenous peoples have also developed methods to assist with hunting. Kite fishing and herbal fish poisons are used in the Pacific Islands, and elaborate fish nets and traps are common in many parts of the world.

FIGURE 1.12 Growing rice and fish together in China has good environmental outcomes.

FIGURE 1.13 The shape of a boomerang allows it to spin, which causes it to return when it is thrown.

FIGURE 1.14 This fisherman in Myanmar is using a traditional fish-trap net.

> **REVIEW**
>
> 1 Explain the expression 'Mother Earth' as used in many Indigenous cultures.
> 2 Give a summary of a creation-time story or myth.
> 3 What is meant by the expression 'hunter-gatherer'?
> 4 Describe the science of how a boomerang works.
> 5 a Describe what North American Indigenous peoples mean by the expression 'the three sisters'.
>   b What advantages are there in growing the three crops together? Do modern farmers use similar methods?
> 6 Beans are legumes that provide nitrogen to the soil. Explain:
>   a why this is so useful
>   b how legumes are able to do this.

Indigenous knowledge and diet

The Tlingit are an Indigenous people of the north–west coast of North America. A Tlingit expression is 'when the tide goes out, the table is set'. Salmon is a main component of their traditional diet. Herrings, herring eggs, halibut and hooligan are also important parts of the Tlingit seafood diet. Tlingit also hunt deer and goats, and use canoes to hunt seals, sea lions and sea otters.

A balanced diet

The traditional Tlingit diet includes plenty of protein from fish, but the Tlingit also need to ensure they have adequate **vitamins** and **minerals**. The Tlingit possess sophisticated knowledge about how to maintain a balanced diet. They recognise that vitamins are important for almost all bodily processes (including the immune, hormonal and nervous systems). Humans cannot make vitamins so they must be obtained from food. The Tlingit use:
- fish bones as a source of calcium, iron and vitamin D
- fish livers as a source of vitamin A
- fish intestines as a source of vitamins E and B
- wild berries as a source of vitamin C.

There are 13 vitamins necessary for human health. They can be classified as either water soluble (C and B group) or fat soluble (A, D, E and K). Fat-soluble vitamins are absorbed from the intestine into the bloodstream. Vitamins A, D and K are stored in the liver and vitamin E is distributed throughout the body's fatty tissues.

FIGURE 1.15 Freshly caught salmon is a main part of the Tlingit diet.

Water-soluble vitamins (vitamin C and the B group vitamins) are stored in the body for only a short period of time before they are removed by the kidneys. Hence, they need to be eaten daily.

The traditional Tlingit methods of trapping salmon are sustainable. They catch only what they need to feed themselves and do not overfish. Modern commercial methods can harm salmon runs and damage spawning populations. Salmon runs are under threat from many activities, such as dam building, irrigation systems, overfishing, agricultural pollution, and the genetic impact and diseases associated with modern salmon hatcheries.

Catching and preserving food

The Tlingit have developed sophisticated methods to harvest salmon. This harvest usually involves a high level of cooperation between the men, women and children. They build systems of walls and weirs, and drive the salmon into closed areas where they can be easily speared.

In the late 19th century, the Tlingit developed the fish wheel, which they used to capture fish as the fish came up the river. Fish wheels make use of the **kinetic energy** (movement) of the river water to make a wheel go around. This rotating wheel can lift up fish that are swimming upstream so they are trapped in a container above the wheel.

The salmon were cooked over fire or preserved by freezing, or drying and then smoking (Figure 1.17). Drying was done outside and the Tlingit had to be careful that the fish were not eaten by bears or birds.

For the smoking process, the Tlingit split the salmon fillets and slashed them to increase the surface area. The pattern of the slashes could be used to identify the person who carried out the smoking. The smoked salmon was sealed in boxes containing seal oil to protect the food from mould and bacteria. The chemicals in the smoke from the burning wood helped preserve the fish and gave it an agreeable taste. Smoking is now a common way of preserving food worldwide.

Go to http://mypsci3.nelsonnet.com.au and click on **Salmon boy** to read the Tlingit myth about a boy who learnt to show respect to the salmon so the fish would return each year to feed the people. What is the importance of myths such as this?

FIGURE 1.16 A fish wheel on a salmon river

Health advantages of 'eating local'

As shown by the Tlingit example, Indigenous peoples have been 'eating locally' for thousands of years. Their knowledge about diet and good health is very sophisticated. One of the problems Indigenous people have when they abandon their traditional diets for more modern processed foods is that their health can suffer enormously. In Canada, Indigenous peoples have much higher rates of obesity and around 4.5 times higher rates of diabetes than the general population. On Pohnpei Island in Micronesia, concern about the health effects of imported food led to the campaign Let's Go Local with an emphasis on educating people about the importance of following traditional local diets.

FIGURE 1.17 The Tlingit method of drying fish prior to smoking

Buying locally

It is interesting that in recent years, people have returned to the idea of buying food grown locally, rather than depending on food transported large distances. Some cities have established community gardens for apartment dwellers, as well as issuing permits for small backyard chicken coops for egg production. This idea reflects many people's desire to live a more sustainable lifestyle.

TA BUY LOCAL

Next time you're buying groceries, consider where the food has come from. What food is grown or produced in your local area? What food is brought in from much further away? Discuss with a partner the top 10 reasons for buying locally, and make a flyer to distribute to fellow students.

Go to http://mypsci3.nelsonnet.com.au and click on **Sustainable food production** to learn more about the movement to eat local food.

Cooking with heated stones

Almost all Polynesian Indigenous cultures have developed methods of cooking with hot stones. A traditional New Zealand Maori *hangi* is laid by first digging a pit in the ground. Suitable stones are heated in a fire and then placed in the pit. Baskets of food are put on top of the hot stones and everything is covered with leaves, wet cloth and then soil. The food is cooked slowly (over about four hours) by the steam produced. A variety of meats, and vegetables such as kumara (sweet potato), pumpkin, carrots and cabbage, can be cooked in this way.

Igneous rocks are the most effective in this cooking method because they have quite high **specific heat capacities**. This means it takes a lot of energy input (from a fire) to heat the stones. The stones later release this heat slowly, to cook the food in the *hangi*.

The energy released by the hot stones heats the water from the wet cloths, forming steam. Steam contains a lot more energy than water so the food cooks more quickly. The soil on top traps the steam and the pressure inside the *hangi* rises. The *hangi* works in a similar way to a pressure cooker.

Kalua is a similar Hawaiian method of cooking. The underground oven is called an *imu*. Very hot volcanic rocks are used and it is common to cook pig covered in banana leaves. The Indigenous peoples in the Chiloe Archipelago of Chile use a similar method of cooking called *curanto*. Some people believe this supports the theory that there was contact between American and Polynesian peoples in the past.

FIGURE 1.18 A Maori *hangi* cooks with steam.

FIGURE 1.19 *Curanto* involves cooking over hot stones on Chiloe Island in Chile.

Cooking with hot stones

INVESTIGATION 1.1

YOUR CHALLENGE
To design an investigation that compares the efficiency of different stones for use in a Maori *hangi*, a Hawaiian *kalua* or a *curanto* of the Chiloe Archipelago.

THIS MIGHT HELP
Some substances when heated to a certain temperature hold much more heat energy than other materials. This property is referred to as 'specific heat capacity'. The higher the specific heat capacity of a stone, the more effective it will be as a source of heat in a steam oven.

A basic method is to first heat a stone to quite a high temperature in hot water, then put the stone into cold water and measure the temperature change of the water. You could investigate variables such as the type and size of the stone. How will you make it a fair test? You could also investigate the speed at which the cold water is heated by the stone.

Carry out and write up your investigation following the guide in Appendix 3 on page 177 or as advised by your teacher.

SAFETY
Do not heat the stones in a direct flame, as they can shatter. Use tongs to handle the hot stones.

CRITICAL THINKING
Designing scientific investigations: consider how to write an effective research question.

REVIEW

1. Outline the diet of the Tlingit people. Discuss whether it is a healthy diet.
2. Find out about the traditional diet of the Indigenous people in your area.
3. Discuss why it is important to get vitamin C in your diet every day. How does the body treat vitamin A and vitamin C differently?
4. Outline how the Tlingit catch and preserve salmon.
5. Discuss why many salmon runs are under threat.
6. Explain the science of how hot stones can be used for cooking, such as in the Maori *hangi*.
7. Outline what we mean by the 'specific heat capacity' of a substance.
8. Which would be better for an oven that uses hot stones for cooking: igneous rocks or sedimentary rocks? Explain your answer.

Go to http://mypsci3.nelsonnet.com.au and click on **Tlingit** for more information on the Tlingit people.

Indigenous knowledge about medicines

Indigenous peoples knew many cures for illnesses long before medicine in Europe became a science. Table 1.1 summarises some of these Indigenous remedies.

TABLE 1.1 Traditional Indigenous remedies

Indigenous group	Remedy	Modern scientific explanation
North American people	They cured goitre by eating harvested sea kelp.	Sea kelp contains iodine, which prevents goitre.
The Hurons in Canada	They cured scurvy in the 1500s by boiling the needles from an evergreen tree then drinking the water.	The needles have high concentrations of vitamin C.
Indigenous peoples in North America	They use 'sacred bark' from the California buckthorn tree to clean out the bowel in cases of constipation.	'Sacred bark' extract is now the basis of some laxatives produced by pharmaceutical companies.
Aboriginal people in south-eastern Australia	They treat sore throats and skin infections with emu bush.	The bush has good antibacterial properties.
Many Indigenous peoples in Africa	They use a plant called *Securidaca longipedunculata* to treat wounds, sores and coughs. It is also used as a pesticide.	The plant has good antibacterial properties, and a substance known to work as a pesticide.

FIGURE 1.20 Bearded tongue is used by Indigenous North Americans for healing wounds, stings and sunburn.

FIGURE 1.21 The qinghao plant, the source of artemisinin, an antimalarial drug, was originally used in traditional Chinese medicine.

FIGURE 1.22 The chamomile plant and flower has a long history of use by Indigenous peoples.

FIGURE 1.23 African apes eat certain plants that get rid of parasites in their intestines.

Tribal people in north-east India use plants to treat fevers, bronchitis, blood and skin diseases, eye infections, ulcers, diabetes and high blood pressure. The knowledge of the plants and their uses is passed on by the *vaiyas*, Indian herbal medicine doctors.

Indigenous peoples can be seen as the custodians of the world's genetic plant heritage. Nearly a quarter of all modern medicines are derived from natural products, many of which were first used in traditional remedies.

Chamomile: a traditional herbal medicine

Chamomile has a long history as a medicine in various parts of the world. The active constituents of chamomile have anti-inflammatory properties, and ease spasm and discomfort in the digestive tract. Other uses include the treatment of anxiety, insomnia, fever, eye irritations, diarrhoea, eczema, menstrual disorders, irritable bowel syndrome and skin irritations. It is also thought to slow fungal growth.

Chamomile is thought to help the growth and health of other plants, especially those that produce essential oils. It is thought to increase production of those oils, making certain herbs, such as basil, stronger in scent and flavour.

Indigenous healers often claim to have learnt by observing sick animals. At times, the animals would change their food preferences to nibble at bitter herbs they would normally reject. Sick animals tend to eat plants rich in substances that have antiviral, antibacterial or antifungal properties.

Properties of chamomile

INVESTIGATION 1.2

YOUR CHALLENGE
To investigate one of the following questions.
- Does chamomile have antifungal properties?
- Does chamomile encourage the growth and/or oil production of other plants?

THIS MIGHT HELP
You could make chamomile solutions from commercially available chamomile tea preparations. Will you use a variety of concentrations?

Your teacher will help you grow some suitable samples of fungi. You could use slices of bread in closed plastic packages.

Carry out and write up your investigation following the guide in Appendix 3 on page 177 or as advised by your teacher.

SAFETY
- Moulds can cause allergies and infections. You should wear gloves and masks during this investigation.
- Do not open the plastic packages once the mould has started to grow. At the end of the experiment, the school will need to dispose of the plastic packages following local requirements.

The health of Indigenous people

Indigenous peoples have experienced serious health problems as a result of their encounters with their colonisers. Exposure of these previously remote Indigenous populations to European diseases caused many fatalities. The local people had previously not been exposed to these diseases and so had no natural resistance to them. For example, the Spanish conquistadors in the 15th and 16th centuries brought smallpox, chicken pox and measles with them to South America (Figure 1.24). Historians estimate up to 85% of the Indigenous population in South America was killed as a result of these diseases.

In modern times, many Indigenous peoples are experiencing a variety of health problems. This can be caused by poor diet, the disappearance of local knowledge, loss of their traditional homelands, environmental contamination and climate change. For instance, Indigenous people have a much higher rate of tuberculosis. Among the Guarani people in Bolivia it is 5–8 times higher than in the general population, and in the Kalaallit Nunaat in Greenland it is 45 times higher.

FIGURE 1.24 Spanish conquistadors brought new diseases to Indigenous peoples.

Some other health statistics from the World Health Organization in relation to Indigenous people include:

>Diabetes: In some regions of Australia, the Aboriginal and Torres Strait Islanders have a diabetes prevalence rate as high as 26%, which is six times higher than in the general population.
>
>Living conditions: In Rwandan Twa households, the prevalence of poor sanitation and lack of safe, potable water were respectively seven times and two times higher than for the national population.
>
>Reproductive health: For ethnic minorities in Viet Nam, more than 60% of childbirths take place without prenatal care compared to 30% for the Kinh population, Viet Nam's ethnic majority.
>
>Suicide: Among Inuit youth in Canada, suicide rates are among the highest in the world, at eleven times the national average.
>
>Infant mortality: Average infant mortality among Indigenous children in Panama is over three times higher than that of the overall population (60–85 deaths per 1000 live births versus the national average of 17.6).

© Copyright World Health Organization (WHO), 2015. All Rights Reserved.

Incidence of infant mortality in Indigenous cultures

ACTIVITY

Consider the data shown in Figure 1.25.
1. What does this data show you about the differences between the incidence of infant mortality in Indigenous people and the population as a whole in these countries? What reason do you see for these differences?
2. Comment on the differences between countries. What reason do you see for these differences?

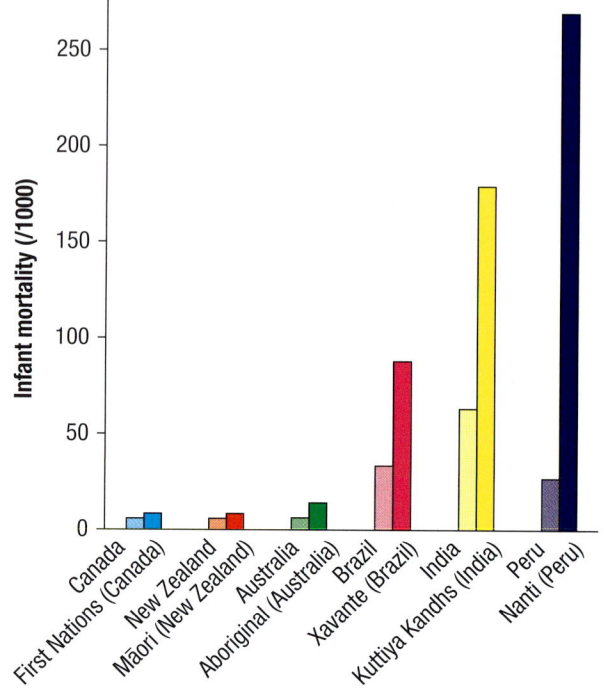

FIGURE 1.25 Incidence of infant mortality

RESEARCH
Evaluating information from the internet critically

Efforts to improve Indigenous health

ACTIVITY

RESEARCH

1. Carry out further research into efforts being made by many organisations to improve the health of Indigenous people worldwide.
2. Summarise what you feel are the major health-related issues facing Indigenous peoples.
3. Imagine you were in charge of the World Health Organization. What recommendations would you make to help improve the health of Indigenous people? Make sure you show understanding that simply making more modern medical care available is not the full solution.

Indigenous peoples and intellectual rights

Consider who owns (has the **intellectual rights** to) the knowledge that Indigenous peoples have of plants and their environment. Article 31 of the United Nations Declaration on the Rights of Indigenous Peoples (2007) states that:

> Indigenous peoples have the right to maintain, control, protect and develop their cultural heritage, traditional knowledge and traditional cultural expressions, as well as the manifestations of their sciences, technologies and cultures, including human and genetic resources, seeds, medicines, knowledge of the properties of fauna and flora, oral traditions, literatures, designs, sports and traditional games and visual and performing arts.

They also have the right to maintain, control, protect and develop their intellectual property over such cultural heritage, traditional knowledge, and traditional cultural expressions.

UN Declaration on Rights of Indigenous Peoples (UNDRIP) related to Sustainable Development and Environmental Change, Article 31. © UNESCO http://www.unesco.org/new/en/indigenous-peoples/sustainable-development-and-environmental-change/undrip-sd-ec

There are many different opinions on whether it should be possible for other people to **patent** the genetic information in plants used by Indigenous peoples.

The neem tree

Some people question the right of pharmaceutical companies to make large profits from medicines developed using Indigenous peoples' knowledge without their permission. This is sometimes called **biopiracy**.

The neem tree, which grows throughout India, has many traditional medical uses. Pharmaceutical companies have shown interest in producing modern medicines from its ingredients. In 1995, the US government awarded a patent to a pharmaceutical research company relating to a method of extraction of an antifungal compound from the tree. One of the conditions for the award of a patent is that it is based on new knowledge. The pharmaceutical company had argued that as the Indian knowledge about the neem tree had never been published in an academic journal, it was not proper knowledge.

This caused an enormous reaction within India. Eventually, after legal action by the Indian government, the patent was withdrawn in 2005. From cases like this, the Indian government has started a major project to document all available local knowledge of the medicinal uses of plants.

FIGURE 1.26 A neem tree (*Azadirachta indica*) has many uses in traditional Indian medicine.

Other recent examples include the successful marketing of an appetite suppressant to help people lose weight, based on chemicals in a succulent plant found in South Africa.

> ### REVIEW
>
> 1. Explain why the local Indigenous populations of South America were very vulnerable to diseases brought by the Spanish conquistadors.
> 2. Many Indigenous peoples experience general health problems when they adopt more modern lifestyles. Why might this be? What solutions do you see?
> 3. Give two examples of Indigenous herbal medicines that have been developed by pharmaceutical companies.
> 4. Describe your thoughts on the idea of pharmaceutical companies being allowed to take out patents on products based on the traditional herbal medicines of Indigenous peoples.
> 5. a Find out about some other medicines that have their origins in traditional Indigenous remedies.
>   b Research the history of aspirin.

Go to http://mypsci3.nelsonnet.com.au and click on **Indigenous medicine** to discover more about the Indian government's project to document knowledge of medicinal uses of plants.

Indigenous knowledge about navigation

Go to http://mypsci3.nelsonnet.com.au and click on **Songlines** for more information on Australian Aborigines' use of the stars in their songlines.

Go to http://mypsci3.nelsonnet.com.au and click on **Navigation** to find out more about latitude and longitude.

Polynesians are well known for their navigation skills, developed during their voyages over the past 2000 years. During the day, they used knowledge of the Sun and the wave patterns caused by nearby islands. At night they used a sophisticated knowledge of the stars and planets to be sure of their direction. From the position of the stars, they could tell both their latitude and longitude. They had memorised the positions over the year of major constellations of stars such as the Pleiades, which they called *Makalii*, and the movement of planets such as Jupiter, which they called *Iao*, and the red planet (Mars), which they called *Hoku-ula*.

FIGURE 1.27 A Polynesian boat as observed in 1616 by Dutch sailors

FIGURE 1.28 The Pleiades (or Seven Sisters), a cluster of stars that is often used for navigation

The Kayapo: A modern case study

The Kayapo is an Indigenous group of people who live in the Brazilian rainforests, on the Xingu River. They use more than 650 different plants for medicinal purposes. They hunt and fish for their food, and carry out shifting cultivation. This means they are nomadic – they farm a certain area for a few years and then move to another area. This allows the old land to lie fallow and replenish its **nutrients** naturally.

This idea of leaving land to lie fallow has always been considered good farming practice. Modern intensive methods of agriculture usually don't follow this practice and instead replace lost nutrients with artificial fertilisers. This can lead to reduced soil quality and sometimes soil erosion.

The Kayapo people, and other Indigenous tribes of the Amazon rainforests, are under threat in many ways. Their land is being taken from them for agriculture and mining, often gold mining and petroleum. Pollution, particularly of their rivers, is also a problem.

A particular issue facing the Kayapo is that the Brazilian government plans to establish dams along the Xingu River. The Kayapo have been very effective in making their situation known to the world. They have protected their land and culture against the interests of people wanting their land for agriculture, logging and mining. For a while, their campaign stopped the Brazilian

FIGURE 1.29 The Kayapo people from the Amazon region live a traditional lifestyle, which is under threat.

government from damming the Xingu River, but this was short-lived. The world's third-largest dam is now being built on the Xingu River and the Kayapo face losing their ancestral homelands and way of life. They have been offered a large amount of financial compensation but this won't make up for the breakdown in their traditional way of life.

The Kayapo cleverly used modern technologies such as video and TV to tell their story. The rock star Sting made a highly publicised appearance at one of their demonstrations (Figure 1.30).

The Kayapo are also well known for their relationship with The Body Shop. In the early 1990s, the Kayapo entered into a relationship with The Body Shop to sell Brazil nut oil, which was used as an ingredient in cosmetics. This initiative was part of The Body Shop's 'Community Trade' projects, which encouraged fair trade with Indigenous peoples to make use of their local resources to generate an income. The Kayapo villages were paid above market rates for the Brazil nuts. However, some people, including some Kayapo themselves, have argued that the Kayapo were not well treated, and that adequate compensation was not paid for the use of their images in the advertising.

It is a difficult issue. Many people believe that Indigenous peoples such as the Kayapo should not become involved in trade with such large international companies.

FIGURE 1.30 The singer Sting appeared at a demonstration supporting the Kayapo in their fight against dams being built along the Xingu River.

Go to http://mypsci3.nelsonnet.com.au and click on **Xingu River** for a video of the Xingu River hydroelectric power station project.

Fair trade debate
ACTIVITY

Research other examples of fair trade. Prepare for a debate with your class, with half the class considering issues from an Indigenous group's perspective and the other half from the perspective of companies from industries such as cosmetics and food.

Some films to watch
ACTIVITY

CHILDREN OF THE AMAZON (2008)
This film of the Surui people was made by Brazilian filmmaker Denise Zmekhol. The film describes what happened to the Surui's life in the largest forest on Earth when a road was built straight through its heart.

AVATAR (2009)
James Cameron, the creator of *Avatar*, was recently on a panel with a number of Indigenous North Americans discussing Indigenous issues. Why would this be? *Avatar* deals with a fictional tribe of humanoid creatures called the Na'vi, who inhabit the rainforest world of Pandora. The Na'vi must fight to preserve the forest from a mining corporation backed up by military force.

CLASS DISCUSSION
1. As a class, discuss how *Avatar* parallels the situation of the Indigenous peoples in the Amazon rainforests.
2. What message did the films create?
3. Write and share a personal response to *Children of the Amazon*, perhaps a poem or a declaration speech.

Impact of modern conservation approaches on Indigenous people

> Our great grandparents lived with the animals, and took great care of them. But we are being chased out from them. We need to be given our land. Now it is very difficult for us to survive.
>
> Matsipane Mosetlhonyana, a bushman who was evicted from the Central Kalahari Game Reserve

A challenging issue for conservation worldwide is the rights of Indigenous people to their ancestral lands. The quote above reflects the issue faced in the Central Kalahari Game Reserve in Botswana. On one hand, it is good to see the efforts being made to protect sensitive ecosystems and endangered animal species. On the other hand, Indigenous people argue that they have lived in harmony with their ancestral lands for centuries and are skilled in managing animal populations. It can be argued they are part of the ecosystem.

The Botswana Bushmen (Figure 1.31) are being forced off their ancestral lands by antipoaching squads. The Botswana Bushmen argue that there is a big difference between the sustainable hunting carried out by Indigenous people and the illegal poaching being carried out for short-term economic gain.

Another seemingly positive step was the Convention on International Trade in Endangered Species (CITES), which banned the ivory trade (from elephant tusks) in 1989. Again this has had the effect of restricting the access of Indigenous peoples to traditional and sustainable hunting. It seems important that local Indigenous peoples are involved in conservation projects. This has happened successfully in the Namunyak Wildlife Conservation Trust, a community-run conservation area that was established in 1995 by the Samburu people in Kenya. The Canadian government has also successfully involved local Indigenous people in the conservation of polar bears.

FIGURE 1.31 Botswana Bushmen are protesting against local conservation policies that punish them for carrying out their traditional hunting practices.

REVIEW

1. Describe how the Polynesian people gained the skills to make long, complicated journeys.
2. State what is meant by 'shifting cultivation'.
3. a Outline the story of how the Kayapo tried to defend their ancestral lands.
   b Describe your feelings about the conflict of interest between a country wanting to improve electricity supply to its people and Indigenous people defending their rights.
4. Explain what is meant by 'fair trade'. What issues does it try to overcome?
6. Explain why conservation projects sometimes cause conflict with the rights of Indigenous peoples.

UNIT QUESTIONS

CRITERION A

EXPLAINING SCIENTIFIC KNOWLEDGE

1. List the main stages in the modern scientific method. (Level 1–2)
2. How is most knowledge communicated in Indigenous cultures? (Level 1–2)
3. State two examples of effective traditional medicines that have modern scientific explanations. (Level 3–4)
4. State three diseases the Spanish conquistadors took with them to South America and the impact these diseases had. (Level 3–4)
5. Outline the importance of creation stories for Indigenous peoples. (Level 5–6)
6. Outline the importance of salmon fishing to the Tlingit people. (Level 5–6)
7. Describe how the development of Indigenous knowledge is different from that of modern scientific knowledge. (Level 7–8)
8. a Describe the different categories of vitamins.
   b Explain why some vitamins, such as vitamins C and B, need to be consumed every day, whereas other vitamins need to be consumed less often. (Level 7–8)

APPLYING SCIENTIFIC KNOWLEDGE AND UNDERSTANDING TO SOLVE A PROBLEM

9. Suggest how Indigenous peoples might have developed their knowledge about medicinal plants. (Level 1–4)
10. Why do you think the health of some Indigenous peoples is worse than that of local non-Indigenous peoples? What solutions do you see? (Level 1–4)
11. Recently the salmon in a river stopped running. What suggestions do you have for why this might have happened? What solutions do you propose? (Level 5–8)
12. Modern agricultural methods and the movement of food around the world often cause a lot of environmental damage. Use your knowledge from this unit to suggest ways this situation can be improved.

INTERPRETING INFORMATION

13. You read the following quotes on a website.

    The environment is not separate from ourselves; we are inside it and it is inside us; we make it and it makes us.

    Davi Kopenawa Yanomami, Yanomami Shaman, Brazil

    Our relationship to the forest is like a child to its mother. The Western environmental groups can't understand that.

    Muthamma, Jenu Kuruba Leader, India

    a Apply what you have learnt in this unit to explain what these quotes are trying to communicate.
    b Give your opinion about whether Indigenous people should remain living in sensitive ecosystems where attempts are being made to conserve endangered species. (Level 1–8)
14. Discuss whether biopiracy is a serious problem or whether all knowledge in the world should be freely available at no cost to anyone. (Level 1–8)
15. An article in a magazine suggests that Indigenous knowledge is poor, unscientific knowledge and has little worth. Write a letter to the editor of the magazine giving your views. (Level 1–8)

REFLECTION

1. How much do you think the culture we grew up in affects our values and the perspectives we have on global issues?
2. Discuss whether you think that people in modern Western societies use the word 'environment' in the same way as Indigenous people.
3. Discuss how we can use the idea of balance in relation to protecting the environment.
4. Do you think it is possible that Indigenous peoples can live in modern societies in a way that maintains their traditions and values?

UNIT 2

ADAPTATIONS OF ORGANISMS

KEY CONCEPT
Systems

RELATED CONCEPTS
Environment

Form

Function

Evolution

GLOBAL CONTEXT
Orientation in space and time – an exploration into how organisms have adapted over time to live in their environments

STATEMENT OF INQUIRY
The adaptations of organisms to changes in their environment can be explained by the theory of natural selection.

INQUIRY QUESTIONS

FACTUAL
1. What is meant by adaptations in organisms?
2. Who was Charles Darwin?

CONCEPTUAL
3. Why is the theory of evolution by natural selection very important to understanding how organisms adapt to their environments?
4. What are the different forms of adaptation, and how are they different?
5. How are humans responsible for the adaptations of organisms?

DEBATABLE
6. Scientists are now able to genetically modify certain organisms for a variety of purposes. Who should be responsible for setting limits on genetic modification, and how should this be done?

Introduction

Organisms have a wide variety of shapes, sizes, behaviours and habitats. During the 19th century, naturalists and explorers first started inquiring into the relationships of living organisms to each other and their environment. The human-built environment of highways and sprawling cities had not yet divided the Earth's natural landscape. Newly introduced **species** had not yet invaded established habitats.

Organisms thrive in certain places because of favourable conditions, where they can reproduce, obtain enough food and water, and find shelter. When conditions change due to natural events or human impact, the organisms must adapt to their changing habitat or die. Many organisms are able to adapt to extreme conditions.

Adaptations to extreme conditions

Research

Work in groups to carry out research into how organisms have adapted to:
- altitude
- chemicals in the environment
- cold
- dry conditions
- fire
- heat.

You will find a useful BBC website for research in the weblink.

Your product

Produce an informative and attractive poster for an online science museum. Your poster should summarise what you have learnt and be suitable for students of your age. Include suitable hyperlinks and acknowledge all sources you have used.

Go to http://mypsci3.nelsonnet.com.au and click on **Adaptations** to help your research on adaptations.

COMMUNICATION
Use of appropriate forms of communication for different purposes

Adaptations

Adaptations are adjustments or changes in the structure or function of an organism in response to conditions in the environment around them. These adaptations give the organism a better ability to compete for resources and hence a better chance to survive. Change in available food and water, change in **climate**, and change in the environment can each have a drastic effect on an organism. Organisms may not grow as large or reproduce as often when circumstances are severe. There are three main types of adaptation organisms can undergo: structural, behavioural and functional. The process of adaptation is explained by Darwin's theory of evolution by natural selection.

The basis of the theory of evolution is as follows.
- Individuals in a species show a wide range of variation in their characteristics.
- This variation is because of differences in their genetic make-up.
- Individuals with characteristics most suited to the environment are more likely to survive and reproduce.
- These successful individuals will pass these characteristics to their offspring (via their genes).

Structural adaptations

Adaptations help organisms survive in their environments. Examples of **physical** (or **structural**) **adaptations** in animals include a giraffe's long neck, a beaver's fatty, oar-like tail, the special sense organs in sharks and types of camouflage. The hair structure of a polar bear's fur helps it stay warm in sub-freezing temperatures. Birds' beaks have adapted to their habitat. Water birds may have webbed feet for paddling or long, spread-out toes for wading (Figure 2.1). Alligators and crocodiles have nostrils that stay above water while the rest of the animal is submerged. Structures such as teeth can tell you a lot about an animal's diet, even one that is long extinct.

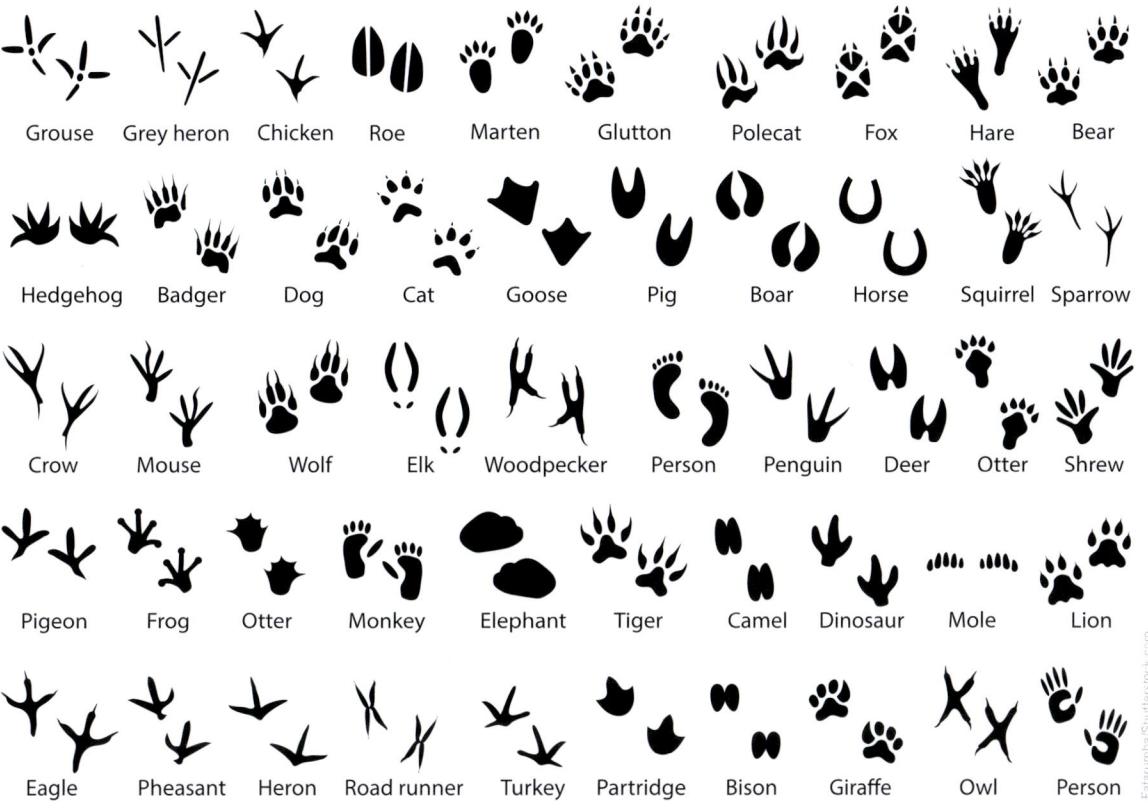

FIGURE 2.1 The structure of animal feet can show adaptations to their unique environments.

The thick wax on the leaves of the mirror bush (*Coprosma repens*, Figure 2.2) is an example of a structural adaptation in a plant. The wax reduces water loss, protects the plant against insect predation, and helps control leaf temperature by reflecting sunlight.

FIGURE 2.2 The glossy leaves of the mirror bush (*Coprosma repens*) keep the plant cool and prevent water loss.

FIGURE 2.3 The queen of the night (*Peniocereus greggii*) opens its flowers at night – a behavioural adaptation. What kind of adaptation is the colour of its flower?

FIGURE 2.4 This young fox will learn the hard way to stay away from skunks after it gets sprayed!

Behavioural adaptations

Organisms can change their behaviour to help survive. This is called a behavioural adaptation. Examples of **behavioural adaptations** in animals include postures (such as sitting very still to avoid detection by predators), spreading out and remaining inactive during hot weather, or huddling in a group to share body warmth in cold weather.

Plants can also modify their behaviour. For example, desert plants such as the queen of the night cactus (Figure 2.3) flower at night because the pollinators are **nocturnal** bats.

Functional adaptations

Functional adaptations generally involve the way the organism works, and include the ability to sweat, lower the rate of cellular reactions to hibernate, or release special chemicals that change the behaviour of others. The defensive smell released by skunks (Figure 2.4) and the fragrance of plants that attracts pollinators are examples of functional adaptations.

REVIEW

1 Very few adaptations are exclusively structural, behavioural or functional. For example, if an organism releases an offensive chemical, it may have a special structure to do so. Draw a Venn diagram of the three types of adaptations and decide where to place each of the following organisms on the basis of the adaptations described.
   a A bombardier beetle can squirt hot acid at potential prey.
   b A stick insect stays very still to avoid detection by predators.
   c A chameleon traps its prey with its extremely long, sticky tongue.
   d The sabre-toothed tiger had massive canine teeth, for a reason that is not known.
   e Snails become dormant during hot seasons by sealing their shells with mucus.
   f Elephants can manipulate tiny objects with their trunks.
   g Snakes move using muscles attached to their ribs.
   h Some species of edible butterflies mimic the appearance of nasty-tasting ones.
   i The formic acid smell of many ants protects them from predation.
   j The bright colour of poison arrow frogs is a warning and reminder to predators that they are inedible.

Adaptations in bears and marsupials

Bears

All bears except polar bears are **omnivores**, meaning they eat both plants and animals. Their skulls and teeth have adapted to this diet. Omnivorous bears have flatter molar teeth than do polar bears, which are **carnivores**. Polar bears (*Ursus maritimus*) have sharp teeth to grasp, tear and chew prey. Adult males can weigh nearly 700 kilograms. Huge fur-covered paws distribute this mass on thin ice, and also provide traction while walking and standing. The polar bear's dense, white, insulating fur provides **camouflage** as it stalks its **prey**, the seals (Figure 2.5). These characteristics are ideal for living in the Arctic climate.

Brown bears (*Ursus arctos*) occur in most parts of the world. They are about the same size as polar bears. Their diet is seasonal. They eat both vegetation (such as nuts, berries, fruit, leaves and roots) and meat from small mammals and fish. Brown bears have shorter snouts and blunter teeth than polar bears, and a distinctive large shoulder hump. Their strong shoulder muscles and sharp claws help to tear logs apart to find food and dig up roots. These are adaptations that help it survive.

Other species of bear can be found in different regions of the Earth. Each species has a unique set of adaptations. The spectacled bear (*Tremarctos ornatus*, Figure 2.6) of South America has a flatter face than the brown bear and was once thought to be a **herbivore**, eating unopened palm leaves, palm nuts, orchids and cacti. Now we know that meat forms about 5% of its diet.

The sloth bear (*Melursus ursinus*) of the Indian subcontinent has a loose, shaggy coat. It eats fruit and its long snout and tongue is ideally suited to eating ants and termites. The sun or honey bear (*Helarctos malayanus*) of tropical South-East Asia is another small omnivore. It has powerful teeth to rip leaves and bark off trees to find bugs. Its thin fur is an adaptation to help it live in tropical forests. The American black bear (*Ursus americanus*) is the most common bear in North America. In spite of its name, this bear can be different colours depending on its habitat. In forests it is black. In arid western North America, it can be a brown or cinnamon colour. In very rare cases it can be white. These different colours are adaptations that help to camouflage it in its environment. Asian black bears (*Ursus thibetanus*) are similar to their American counterparts, but are smaller. They have similar teeth to other omnivores in order to eat both plants and animals.

FIGURE 2.5 Polar bears are highly adapted for their natural environment.

FIGURE 2.6 South American spectacled bears were thought to have been vegetarian. The fictional 'Paddington Bear' in children's literature must have been a spectacled bear – he came from 'darkest Peru'!

FIGURE 2.7 The giant panda's back teeth are relatively flat, which makes them suited to grinding and chewing bamboo.

The giant panda (*Ailuropoda melanoleuca*) is the rarest and perhaps the most recognisable bear (Figure 2.7). With distinctive black and white colouring, it was once classified in the raccoon family. However, genetic (DNA) testing shows pandas are most closely related to bears. Scientists think their colouring helps protect them by allowing them to blend in with the dark rock and snow of western China. Their diet is almost completely bamboo. Their back molar teeth have adapted to be able to grind the fibre of tough bamboo plants.

Although scientists can trace a common **ancestor** for all these bears, each has physical characteristics that are adapted to its unique environment. Their genes (DNA) have changed over millions of years from their original common ancestor to what they are today.

Marsupials

Go to http://mypsci3.nelsonnet.com.au and click on **Diversity and adaptions** to choose from a number of interactive simulations on diversity and adaptations.

The kangaroo (Figure 2.8) is a **marsupial**, which belongs to the **mammal** class of animals. Newborn marsupials are always very tiny. They complete their development in a **pouch**, where they suckle on milk until old enough to survive in the wild. Although all kangaroo species are restricted to Australia and surrounding islands, **fossil** marsupials have been found in Antarctica, and dozens of marsupials, opossums and shrew opossums are native to the Americas.

More than 68 000 years ago, Tasmania was joined to the Australian mainland by a land bridge (Figure 2.9). Humans and other animals were able to travel between Tasmania and the mainland.

FIGURE 2.8 Kangaroos are an iconic Australian marsupial species, but other types of marsupials are found around the world.

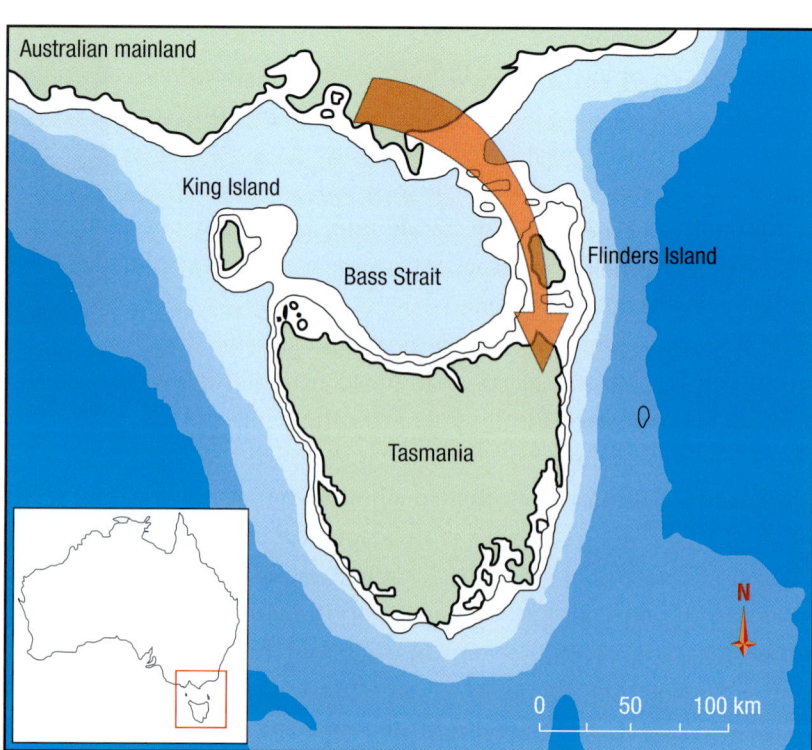

FIGURE 2.9 Map showing a land bridge between Tasmania and mainland Australia, and (inset) Australia today

It is thought that this land bridge survived until about 10 500 years ago. Today, the Tasmanian devil, a carnivorous marsupial, is only found on the island state of Tasmania.

These are just a few case studies looking at the ways some animals have adapted to their environments. They can be repeated for thousands of other organisms, including plants.

Introduced species

Carnivorous marsupials, such as the quoll, used to be more common on mainland Australia, but their numbers have declined. Australia has a very long history of human habitation, but until 4000–6000 years ago, there were no dogs. The dingo (Figure 2.10), which arrived in Australia about 4000 years ago, spread across the Australian continent and competed for the same **niche** in the environment as the native marsupial carnivores. It probably ate the same prey, and might have had a faster **reproductive rate**. Or, as with many **introduced species**, it may have been able to thrive in the new environment because there were no natural checks on the **population**, such as predators and disease. Europeans later introduced other competing species, including cats and dogs.

FIGURE 2.10 The Australian dingo is an introduced species.

Soon after the dingo appeared on the mainland, marsupial carnivores species became much less common or disappeared. This is an example of how a native species, one natural to its environment, is not always better adapted than an introduced **alien species**. Meanwhile in Tasmania, the remaining marsupial carnivores were saved from **extinction** by the barrier of Bass Strait. There are no dingoes in Tasmania to this day.

Go to http://mypsci3.nelsonnet.com.au and click on **Invasive species** to read more about the dangers to human health and economies from invasive species.

Species can be accidentally introduced to a new habitat or intentionally brought in as a response to a problem. The cane toad (*Rhinella marina*, Figure 2.11) is native to Central and South America. It was found to be helpful in eating insects around sugar canes, so it was imported to Hawaii, the Philippines and the Caribbean. Since it was so successful, many other regions, including Australia, imported cane toads. This huge toad with an enormous mouth will eat almost any animal. It secretes a poison through its skin that can sicken or kill small animals. In a new ecosystem, there are no predators to keep its numbers in check and it can outcompete native amphibians for food. Whether introduced accidentally or deliberately, there can be severe consequences when an introduced species becomes out of control.

FIGURE 2.11 The cane toad, once imported to eat pests, has become a big problem in Australia.

> **REVIEW**
>
> 1 List at least two examples of foods these groups may eat.
>   a Carnivore
>   b Omnivore
>   c Herbivore
> 2 Outline how the following features of a polar bear might help it survive in the Arctic.
>   a Sharp teeth
>   c Thick, insulating fur
>   b Wide feet
>   d White colour
> 3 Suggest why, unlike the other bear species discussed, the polar bear is not an omnivore.
> 4 List at least two reasons why the common ancestor of bears may have died out.
> 5 Outline how an introduced species can become a pest.
> 6 Bears and dingoes do not have pouches, as their babies develop in the womb, nurtured through a structure called the **placenta**. Although immature, the babies of placental mammals are generally larger when they are born and much more developed.
>   a Outline why you think there are very few marsupials in the northern hemisphere today.
>   b Give an example of a marsupial and a placental mammal that may have had very similar needs, and competed until one became extinct.
>   c Are there any continents that have both native marsupials and placental mammals?

Evolution by natural selection

FIGURE 2.12 Charles Darwin, 1809–1862

Evolution explains all animal and plant adaptations that help them to survive in their environments.

The variety of specialised adaptations by different organisms may seem astounding. In the 19th century, Charles Darwin, a British naturalist, developed the theory of **evolution** by **natural selection**. In the 1830s, he sailed around the world in a ship called the *Beagle*. After studying many plants and animals in many countries, he concluded that the various species of animals had come into being because of slow and gradual evolution. In 1895, he published his famous book *On the Origin of Species*. This is one of the most important books ever published and has led to a completely different way of understanding nature. Darwin put his ideas forward at a time when most religions taught that all living things had been placed on Earth by God, exactly as we see them now. So Darwin's theory was very controversial, and created a lot of opposition. As evidence to support the theory of evolution, such as fossils and knowledge of genetics, has become stronger, most people and most religions now accept that the theory of evolution does not have to conflict with their religious beliefs. However, some people still see a conflict. You will consider this issue more carefully in the IB Diploma Programme when you study TOK.

Darwin based his theory of evolution on the following set of observations and logic.

Darwin's observations

1 All organisms in a species show variation

All the chicks in a family of birds are different. This happens because each chick inherits its characteristics from both parents. These characteristics are controlled by genes the parents carry. Many combinations of genes are possible, leading to considerable variation in a family but even more so in a species as a whole.

Logic

The birds are under threat from lack of food and predators. The ones who are most likely to survive and breed are the ones that are best adapted to their environment, i.e., best at finding food, mating and escaping from predators.

2 Some characteristics are inherited

For instance, a bird with a particularly strong beak that gives it an advantage when obtaining food can pass this trait on to its offspring.

Logic

Parents who are well adapted for living in their environment are likely to pass on these characteristics to their children. Their children will also be more likely to survive and likely to pass on these characteristics to future generations.

3 We now see the adaptation within the population of birds

After a number of generations, we start seeing more birds with this favourable characteristic, e.g. stronger beaks. We can say this is an adaptation.

Logic

This is a consequence of natural selection.

Evolution caused by natural selection happens over quite long periods of time. Table 2.1 shows how camels have evolved over millions of years. Scientists know this from observing fossils. If there is a rapid change to the environment, then an organism may not be able to evolve quickly enough. It may die out and become extinct.

TABLE 2.1 How camels evolved to their present form

Age	Palaeocene 65 million years ago	Eocene 54 million years ago	Oligocene 33 million years ago	Miocene 23 million years ago	Present
Organism					
Skull and teeth					
Limb bones					

Mutations

Sometimes in nature, **mutations** happen, whereby the gene in an individual is altered, for example, by radiation from the Sun, by a chemical in the environment, or by chance in the body. Most mutations are not useful and disappear from the population. But occasionally they produce a favoured characteristic and accelerate the process of evolution. This process is shown in Figure 2.13.

FIGURE 2.13 This diagram of the evolutionary tree of life shows the consequences of genetic mutation over billions of years.

The Galapagos finches

Although evolutionary changes usually occur over very long periods of time, it is possible to observe them occurring in the wild. For 40 years a British couple, Peter and Rosemary Grant, ran an experiment on Daphne Major, one of the Galapagos Islands. Every year they carefully measured and recorded the beak length of every newly hatched finch – a task involving thousands and thousands of birds (Figure 2.14). During this long experiment, there was a terrible drought. In 1977, 85% of the finches died. The Grants observed that the beaks of the surviving finches were slightly longer than average for this species. This was a small structural adaptation, and it persisted for many generations. During the drought, there were fewer small seeds available. Finches with longer beaks could source alternative food such as bigger seeds, and so they survived to pass on their genes to the next generation of finches.

In 1983, the climate suddenly reversed. Excellent rains ensured lush vegetation and ample seeds. This time, even the occasional baby finches with short beaks survived. Under these conditions, the small difference in beak size may even have been an advantage, as beaks are 'costly' structures to grow, and finches with smaller beaks could invest their energies in other ways, such as having more offspring.

The research and observations made by the Grants demonstrated that species can respond quickly to environmental change. What may be inferred is that during sustained events, such as climate change over millennia, the evolution of a new species is possible.

FIGURE 2.14 Galapagos finches studied by Peter and Rosemary Grant

Go to http://mypsci3.nelsonnet.com.au and click on **Arthropods** to learn more about the evolution of arthropods.

Artificial selection

Domestic animals and some plants have been selected artificially over many thousands of years. Dogs such as the labradoodle (a cross between a labrador and a poodle) (Figure 2.15) were bred in order to provide guide dogs for people with allergies to dog fur. Ranchers and farmers have used **artificial selection** for thousands of years. Dogs that were more docile were bred from more often than those that were not. Cows are bred to produce more milk and chickens are bred to have larger breasts. Broccoli and brussels sprouts were produced through selective breeding from wild mustard plants. Artificial selection can have positive outcomes, such as increased food production.

Genetic modification

FIGURE 2.15 Dogs such as the labradoodle have been artificially bred for a specific purpose.

Scientists are now able to do more than just cross-breed. They are able to directly modify the genes (DNA) of an organism. This introduces traits to the organism that would not naturally occur. Scientists may use genetic information (DNA) from bacteria, viruses, insects, animals or plants. For instance, they can make certain plants more disease resistant, reduce their water needs or increase crop yield. These are helpful characteristics to have in areas of poverty and hunger.

However, genetic modification is controversial. It is a relatively new technology, and so the long-term effects are unknown. Scientists can only predict what the outcome will be in the new organism. Many people do not want to eat genetically modified food for this reason. There is wide agreement among the scientific community that genetically modifying humans is unethical and it is not permitted. There are significant benefits and possible devastating consequences to genetic modification. Being informed of all aspects of this will be key for future generations grappling with this difficult question.

REVIEW

1. Outline the process of evolution.
2. Describe how Peter and Rosemary Grant identified evidence for evolution of finches in the Galapagos Islands.
3. Outline reasons scientists may think the research done by the Grants is valid.
4. Outline how animals can aid in the spread of other organisms, such as plants.
5. Outline how a farmer might use artificial selection to improve a herd of animals.
6. Outline how you think dog breeders use artificial selection to produce purebred animals with certain characteristics.
7. Predict and explain a danger of genetically modifying an organism.

Plant adaptations to dry environments

Plants take in carbon dioxide from the air and water from the soil and use light energy to produce sugar and release oxygen. This reaction is called **photosynthesis** and it occurs inside the chloroplasts of plant cells. It is the green pigment inside the chloroplasts, **chlorophyll**, that is able to absorb the Sun's light energy. Carbon dioxide from the air enters the plant through very small pores known as **stomata** (singular stoma) (Figure 2.16), generally found on the leaves.

There is a cost to the plant in allowing useful gases to enter the leaf. When the stomata are open, water molecules can escape to the atmosphere. This water, and the water used in the photosynthesis reaction, was absorbed from the soil by the plant's roots. Although water is easy to obtain for plants like water lilies, for many plants it is obtained at great cost. For this reason, many plants have developed adaptations to minimise their water loss.

Most **xerophytic** plants (desert plants such as cacti) minimise their leaf surface area and photosynthesise through their stems. The cost of decreasing their leaf surface area is that xerophytes can only grow slowly. Other plants solve the problem by positioning most of their stomata on the bottom, shaded surface of the leaf. Sometimes the leaf has a waxy **cuticle**, or the stomata may be sunken slightly below the leaf surface. Leaf hairs around the stomata may trap the slightly humid air that is escaping and so decrease the amount of evaporation. Grasses may roll up their leaves in the heat, and many plants, including eucalypts and acacias, have vertically hanging leaves that avoid exposure to the sun during the hottest part of the day. In this way, plants adapt to their environments.

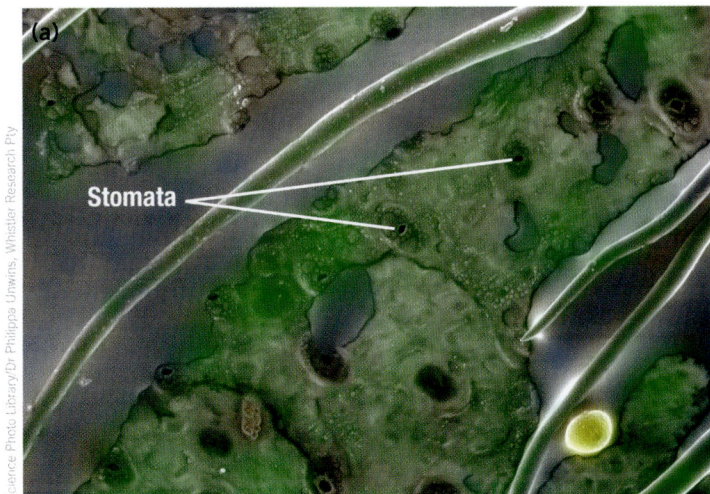
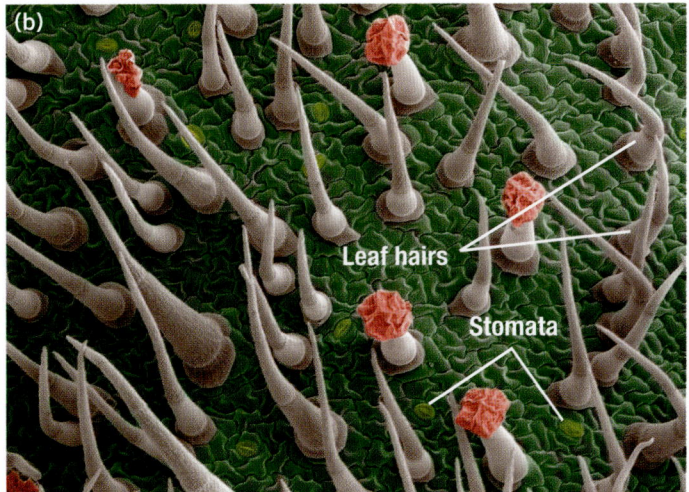

FIGURE 2.16 Scanning electron micrographs of (a) a leaf surface showing sunken stomata, an adaptation to drought, and (b) the surface of a tomato leaf, showing leaf hairs

PERFORMANCE ASSESSMENT TASK — CRITERION C (I AND II)

Observing stomata

EXPERIMENT 2.1

AIM
To compare the stomata of different types of leaves.

MATERIALS
- leaves of a variety of plants – Asiatic dayflower (*Commelina communis*), lettuce, and geranium work well.
- microscope (USB microscope if you have one)
- microscope slide
- cover slip
- tweezers
- pipette

PROCEDURE
1. Carefully rip and peel the top part of the leaf from the bottom. Look for a clear membrane from the bottom layer of the leaf.
2. When you have obtained a small piece of this membrane, use the tweezers to peel it away from the leaf.
3. Carefully place the membrane into a water drop on a microscope slide. Be careful not to fold the membrane.
4. Place the cover slip over the specimen and observe the membrane on low and then high power of your microscope.

RESULTS
1. Very carefully draw and label one stoma. Note the total magnification.
2. Estimate the number of stomata you can see.
3. Compare the results from the different plants.

CONCLUSION
Write a summary of what you have learnt from carrying out this experiment.

ATL — COLLABORATION
Roles in groups. As a scientist, it is very important to understand the importance of collaboration. When performing experiments, help each other by discussing what you are thinking and doing. Scientists will always work in a team to share ideas and collaborate.

PERFORMANCE ASSESSMENT TASK — CRITERIA B AND C

Controlling water loss from leaves

INVESTIGATION 2.1

YOUR CHALLENGE
Design an experiment that will allow you to investigate the variables that control how much water is lost through plant leaves.

THIS MIGHT HELP
You could investigate a number of different plant types from different environments to determine whether water loss varies in different species. Alternatively, you could investigate the different variables that affect the rate of water loss.

Carry out and write up your investigation following the guide in Appendix 3 on page 177 or as advised by your teacher.

ATL — CRITICAL THINKING
Controlling variables is an essential element of any scientific investigation. In this investigation, consider carefully how well you have controlled variables.

REVIEW

1 Identify and discuss the costs and benefits of the following plant adaptations.
   a Mangroves are able to survive in sea water and oxygen-poor silt. Glands on their leaves excrete excess salt, and extensions called **pneumatophores** bring oxygen to their root systems. Mangroves grow very slowly.
   b Mistletoes (Figure 2.17) use photosynthesis and keep their stomata open day and night. Mistletoes are **parasites** and, instead of having roots, they tap into the water vessels of the host plant. Eventually the host plant may even die. Mistletoes depend on small birds to transfer their sticky fruit to new host trees.
   c Plants that live on the floor of tropical rainforests (Figure 2.18) may not get a lot of exposure to sunlight. Photosynthesis occurs in the chloroplasts, which contain the green pigment chlorophyll. The greener the leaf, the more it can photosynthesise in low light. Forest floor plants are generally very dark green in colour and very slow growing.

FIGURE 2.17 Mistletoes use their host trees for water for photosynthesis. This parasitic relationship can damage or kill the tree.

FIGURE 2.18 Forest floor plants are generally dark green and very slow growing.

2 In spring, as sap rises in the rose bush and it starts budding, aphids (also called greenflies) begin to multiply. Aphids suck the sweet sap and can reproduce very quickly. On some rose bushes, ants can be seen stroking the aphids (Figure 2.19) with their antennae to extract the 'honeydew', which they eat. On other rose bushes, the spotted larvae of the lady beetle feed on aphids.
   a How many organisms are there in this scenario?
   b Which organisms will there be the most of, and why?

FIGURE 2.19 This carpenter ant will milk its 'herd' of aphids for honeydew.

Human impact on natural communities

With human intervention, some species, such as those we have domesticated, have thrived. For example, there are now more pet dogs and cats than would ever have lived in the wild. Introduced species have also benefited from humans. The success of introduced species, such as sparrows, myna birds, cane toads and eucalypts, in their new environments has often made them pests, as they compete with **native** species. In other cases, humans have affected rare animals more directly, as summarised in Table 2.2.

TABLE 2.2 How humans have caused species extinction

Cause	Description of cause	Examples
Deforestation	Destruction of forest due to logging or fire clearing	Javan tiger Bali tiger
Pollution	Dumping of waste into waterways and oceans, e.g. oil spills	Speckled cormorant Baiji river dolphin
Over harvesting	Removing more than the numbers allowed by fishing quotas	Tecopa pupfish Labrador duck
Culling	Mass killing using poisons, trapping or shooting	Thylacine Caspian tiger
Animal smuggling or poaching	Illegal trade in protected species	Chinese elephant Black rhinoceros

Directly or indirectly, we are all responsible for these extinctions. Although we may choose not to be part of illegal activities and avoid purchasing fish species at risk of being over harvested, we all benefit from resources such as wood and oil. The food we buy today may have been grown on farmland that was previously home to animals such as the carnivorous marsupial the thylacine (Figure 2.20), last observed alive in a Tasmanian zoo in 1936. Like all organisms, we are part of the **ecosystem**.

FIGURE 2.20 Thylacines, such as this one photographed in captivity in 1900, are now extinct.

Humans can also make a positive difference. Governments can protect and endorse nature reserves, and pass laws. Facilities such as zoos also help, as teaching tools for members of the public, by providing research to learn more about rare species, and through breeding programs. Special facilities have been built for captive endangered animals to breed and live safely. In 1961, the World Wildlife Fund was formed in response to conservationist groups' need for money. Their philosophy is to help build a future where human needs are met in harmony with nature. This organisation works with groups around the world to educate and effect change in response to problems of fresh food and water, forest preservation, ocean life protection and climate change.

Individuals can help native wildlife in their own communities. Local populations of wild animals may be limited by the availability of water, food or places in which to breed. Simple activities such as providing water baths, or planting shrubs and vines to provide food for butterflies and birds, can help these animals. In towns and cities, large old trees are felled because of danger from falling branches. But rotting hollows left by fallen branches are used as nesting sites by many small animals such as birds and bats; sometimes several species use these types of nesting site each year. Artificial fauna boxes (Figure 2.21) can provide alternative nesting sites. Using only native plants in your garden is another big way to help.

FIGURE 2.21 Artificial nesting boxes can assist a variety of small species.

TA HELPING NATIVE WILDLIFE IN YOUR LOCAL COMMUNITY

Look for a community project in your local area that focuses on protecting native wildlife. Some examples are looking after a local wood, protecting local woodland from non-native animals, improving a local river or lake, and protecting a local endangered species. Find out about the work of this community group, their history, who is involved, and their success or difficulties. If possible, you and you class could become involved in the project.

REVIEW

1. Discuss whether there are any truly undesirable species. Outline the effect of removing an undesirable species from a **community**.
2. Imagine you could choose to either save a threatened butterfly or save a carnivorous reptile such as a Komodo dragon (Figure 2.22).

FIGURE 2.22 (a) An Apollo butterfly, one of the many endangered butterfly species; (b) a Komodo dragon

   a. State which you would choose.
   b. State which would require fewer resources.
   c. Describe which choice would have a greater impact on the entire community of organisms they live in.
3. Describe how species sometimes become extinct because of human activities.
4. Discuss the responsibilities humans have to the natural world.

UNIT QUESTIONS

CRITERION A

EXPLAINING SCIENTIFIC KNOWLEDGE

1. Name two carnivorous marsupials of Tasmania. (Level 1–2)
2. List at least three ways in which species may become extinct. (Level 1–2)
3. Camouflage is a structural adaptation that allows an organism to hide in its natural environment. List two examples of species that use camouflage, and state how it is effective. (Level 3–4)
4. Why is it better for a plant on the forest floor to have a lot of green pigment in its leaves? (Level 3–4)
5. List two examples of behavioural adaptations seen in plants. Why can it be difficult to determine behavioural adaptations in plants? (Level 5–6)
6. Explain how the change in environmental conditions gave rise to different beak sizes for the finches of the Galapagos Islands. (Level 7–8)
7. Describe two different ways plants minimise water loss. (Level 7–8)

APPLYING SCIENTIFIC KNOWLEDGE AND UNDERSTANDING TO SOLVE A PROBLEM

8. Why is it advantageous for desert animals to live in burrows and feed at night? (Level 1–2)
9. Compare and contrast the different adaptations displayed by the sun bear of South-East Asia and the polar bear of the Arctic. (Level 3–4)
10. How would you explain the idea of evolution to an adult who had little knowledge of science? (Level 5–6)
11. The cuttlefish (Figure 2.23) is a sea animal that can change its colour voluntarily. The hundreds of special pigment cells on each millimetre of skin can expand or contract to control the visibility of red, yellow, brown or black chemicals. Another layer of cells, iridophores, contain chitin, which reflects blue and green light. Cuttlefish use their ability to change colour to communicate, court mates and hunt. Highly developed eyes, eight arms and two feeding tentacles help it capture food. Like octopus and squid, it can release ink to escape predators. You may have found cuttlebones on the beach. In the living animals, these are internal structures that provide buoyancy and shape.
Create a table and list the structural, behavioural and functional adaptations that are shown by cuttlefish. (Level 5–8)

INTERPRETING INFORMATION

12. A friend suggested that you should be very careful before buying a purebred dog. Explain your reaction to this advice. (Level 1–2)
13. One friend says the giant panda is a member of the raccoon family because of its physical characteristics. Another friend disagrees. Explain your thinking about which is correct. (Level 3–4)
14. Humans have developed farm animals and plants for centuries through artificial selection. Describe the positive and negative aspects of this type of selection. (Level 5–6)
15. Budgerigars (parakeets) are rare in the wild. However, the species is far from extinct because millions of them are kept as pets around the world. Should we domesticate endangered animals such as Tasmanian devils and small monkeys, to save them from extinction? Explain why this would or would not work. (Level 5–6)
16. Animal smuggling and poaching can contribute to the extinction of a species. Some countries destroy animals, birds and eggs caught being smuggled. Many licensed breeders of rare birds offer to incubate the eggs and raise the young birds, so they can be returned to their native habitat. Why might governments be reluctant to take up their offer? What do you think? Justify your answer. (Level 7–8)

REFLECTION

1. A major theme in this unit was how organisms need to adapt to survive in changing environments.
   a. Suggest three ways an environment could change that would require the organism to adapt.
   b. How is the word 'environment' used in other subjects?
2. Explain how the concept of form and function is important to the idea of adaptation in animals.
3. The idea of evolution by natural selection is a major concept in science. However, the term 'evolution' is used in many other subjects and situations. Suggest some other ways we use the idea of evolution.
4. Improving the ability of plants to adapt to dry environments is being achieved in a number of ways, including selective breeding and genetic modification. Some people are very concerned about the implications of genetic modification. Discuss your thoughts, and suggest who should decide which genetic modifications to allow and how these decisions should be made.

FIGURE 2.23 Cuttlefish

UNIT 3
LOOKING AFTER OURSELVES

KEY CONCEPT
Systems

RELATED CONCEPTS
Consequences
Models
Form and function
Cause and effect

GLOBAL CONTEXT
Identities and relationships – an exploration into leading a healthy life

STATEMENT OF INQUIRY
Lifestyle choices we make in adolescence can have physical, social and emotional health consequences both immediately and in the future.

INQUIRY QUESTIONS

FACTUAL
1. What are the roles of the endocrine, respiratory and musculoskeletal body systems?
2. What are the main health-related challenges of adolescence?
3. What are hormones and how do they affect the body during adolescence?

CONCEPTUAL
4. How can the misuse and abuse of drugs affect our bodies?
5. How are physical, social and emotional health related?
6. How do brain changes during adolescence affect decision-making?

DEBATABLE
7. Are the adolescent years the most challenging years of our lives?

Introduction

In what ways can you help your body stay in good working order? Understanding your body's systems and their functions and purposes can help you make better decisions throughout your life. You may ask yourself: 'How does what I eat affect my health?', 'What is the value of exercise?', or 'Should my friends and I smoke?' Just as important as your physical health is your social and emotional health. The adolescent years are packed with important decisions. Just as you are trying to become more independent of the adults around you, you may need their answers to important questions. Learning how to live a full healthy life at this age can help you with your adolescence and the rest of your life.

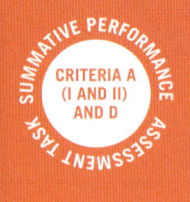

Decisions

Adolescent students face many major lifestyle decisions that can affect their present and future physical and psychological health. You are asked to consider how best to advise adolescents on looking after their health.

Your task

Work as a class to design and produce a health campaign to help students of your age to realise the importance of physical, emotional and social health. First, you will need to consider the elements of a successful campaign. Telling people about the science of good health is usually not enough. How do you design a successful campaign of this kind that will have a deep impact on how people think and act?

If possible, try to run the campaign in your school. The campaign is likely to take place over a reasonable period of time. After you have run the campaign, evaluate how effective it was. This evaluation should also refer to why changing peoples' lifestyles is much more complex than just giving them the necessary scientific knowledge.

ATL

AFFECTIVE

Having perseverance or grit. As you become more independent, you may find yourself in difficult or uncomfortable situations relating to your health. Choosing to resist peer pressure and to keep to your values shows perseverance and grit. These characteristics are important in all aspects of your life, including being a good student. Your health campaign will need to encourage young people to show perseverance or grit.

Adolescents' health – the global challenge

In 2014, the World Health Organization released a report called 'Health for the world's adolescents'. The introduction to this reports states that around one in six people in the world is an adolescent: that's 1.2 billion people aged 10–19.

Most adolescents are healthy, but there is still significant death, illness and disease among adolescents. Illnesses can hinder their ability to grow and develop to their full potential. Alcohol or tobacco use, lack of physical activity, unprotected sex and/or exposure to violence can jeopardise not only their current health, but their health for years to come.

Here is a list of key facts.
- An estimated 1.3 million adolescents died in 2012, mostly from preventable or treatable causes.
- Road traffic injuries were the leading cause of death in 2012, with some 330 adolescents dying every day.
- Other main causes of adolescent deaths include HIV, suicide, lower respiratory infections and interpersonal violence.
- Globally, there were 49 births per 1000 girls aged 15–19 in 2010.
- Half of all mental health disorders in adults appear to have started by age 14, but most cases are undetected and untreated in adolescence.

This is a reminder to us that all around the world adolescents face serious health issues, many specific to adolescents. It is also a reminder that the health risks are very different from one part of the world to another, and from one section of society to another.

You as an adolescent

Around the age of 10–12, you enter a developmental stage called **adolescence**, which will last about 10 years. Your body will go through a series of rapid and sometimes unwelcome changes. It is important to know about the changes that come along with adolescence, and it is even more important to remember that everyone goes through these changes.

What are the changes? You are becoming more independent from your family. You may begin thinking more frequently about your life as an adult. Your body and feelings are changing. As you become older, your parents and community will begin giving you more freedom and responsibilities. It can be a confusing period. Support is available to you and others. As well as your family, there are people in your community who can help you, such as your doctor, the school nurse, teachers and the school counsellor.

Changes during puberty

Puberty is the time when sexual maturity occurs. Many physical and emotional changes happen very quickly. These changes may make you feel uncomfortable and self-conscious. The process of becoming sexually mature can last from one to three years. Not everyone is on the same timetable with puberty, which can add to feeling self-conscious as you go through it. Some begin puberty as early as age 8, others as late as age 15.

During puberty your **endocrine system** begins to flood your body with **hormones**. The endocrine system is made up of eight different glands in the body (Figure 3.1). Hormones are chemical messengers that travel to tissues and organs with information and instructions. They coordinate your growth and development, metabolism, sexual function and mood. They can wreak havoc on your skin (acne) and your emotions (mood swings).

The **pituitary gland**, part of the endocrine system, secretes a hormone called **follicle-stimulating hormone (FSH)** that causes significant physical changes. In girls, hips and shoulders broaden, breasts grow and hair appears in the **pubic** area. Boys grow taller and stronger, and hair starts to grow in the pubic area and on the chest and face.

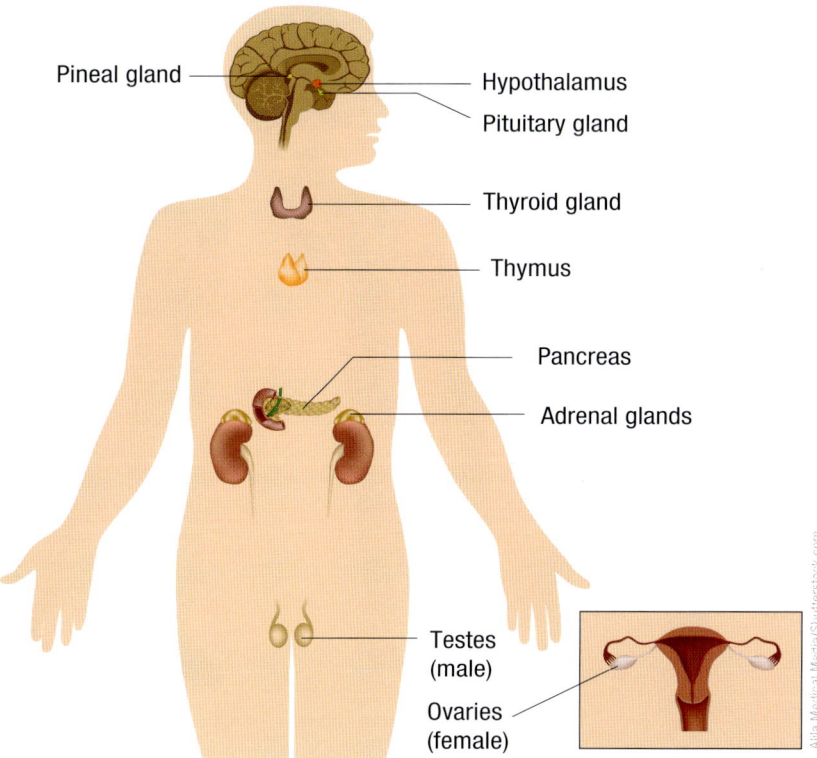

FIGURE 3.1 The eight glands in the endocrine system secrete hormones into the bloodstream.

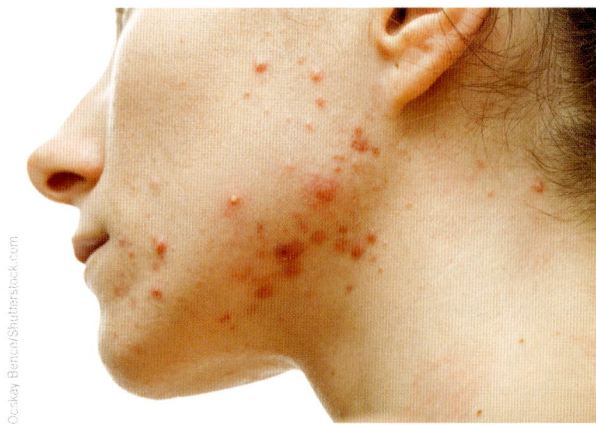

FIGURE 3.2 Acne can be a nuisance during puberty for males and females.

Both boys and girls can get oily skin and develop pimples. In addition, your bodies become reproductively mature. Girls begin their monthly **menstrual cycle**, as the body prepares for nurturing a baby. Boys start to produce sperm.

The importance of body image

These physical changes come at a time where you are more aware of your image and those around you. You wonder where you fit in and compare yourself to others. You might begin to worry about your body image. You need to realise that changes to your body during adolescence are normal and happen in different degrees to everyone. The idealised versions of maleness and femaleness in the media are distortions and do not reflect the average person. Sadly, a very high percentage of adolescents are not happy with their bodies, in spite of the fact that their bodies are totally normal. It seems that social media is increasingly exposing young people to images of bikini bodies, 'six-pack abs' and perfect hair. The pressure to look cool has never been higher. Unfortunately, verbal bullying and cyberbullying about weight are too common.

Young people's body images are influenced by a number of factors. These can include:
- family environment and how the family discuss other people's bodies
- attitudes of peers
- advertising and the media's portrayal of bodies
- the influence of the fashion industry
- cultural background
- social media, especially the posting of photographs.

FIGURE 3.3 Barbie and Ken – what influence have they had on body images?

Trying to copy these false depictions can lead some adolescents to **anorexia nervosa** or **bulimia nervosa**. Both these conditions are extremely dangerous and could lead to serious long-term health conditions or even death. A person with anorexia has a fear of gaining weight. They think a lot about food and limit what they eat. They have a body weight below normal, but still think they are overweight. Their problem is psychological and they often starve themselves as a way of feeling in control of their lives. Anorexia is more common in women. It can cause women to miss menstrual periods, and can lead to future fertility problems. Bulimia is eating a lot of food in a short period of time (binge eating) and then trying to get rid of the food by **purging**: by either vomiting or taking laxative pills (to go to the toilet).

FIGURE 3.4 A woman with anorexia

ACTIVITY — Body image in the media

1. Look through popular magazines or online media to document how men and women are portrayed. Choose three male and three female pictures to analyse. Observe their features carefully and create a table similar to the one below comparing their features to the average person's. Then write a paragraph about the messages you think the advertiser or photographer is trying to convey and whether or not it is healthy. Be conscious that these photographs may have been airbrushed or digitally changed in other ways.

	Picture	Description	Compared to average person
Female			
Male			

2. How would you advise a friend who, in spite of having a normal body, was convinced that she or he should lose weight?

Go to http://mypsci3.nelsonnet.com.au and click on **Body image girls** or **Body image boys** for more advice on the issue of adolescents developing poor body images.

Physical health

Regular exercise

Regular exercise is just one component of looking after the physical you. The habits you develop in relation to exercise during adolescence are likely to stay with you throughout your life. As you know from *Science 2 for the international student* Unit 3, your body consists of many finely tuned systems working together. These need to be kept in good shape in order to be healthy.

Regular physical activity has many benefits, particularly during adolescence. It helps build bone and muscle and reduces the risk of **chronic disease** and obesity. Circulation from exercise increases oxygen delivery to the skin, making it healthier. Exercise also promotes feelings of well being and reduces stress and anxiety, which can in turn reduce the risk of depression It is recommended that adolescents do a minimum of 30 minutes' exercise daily.

The human musculoskeletal system

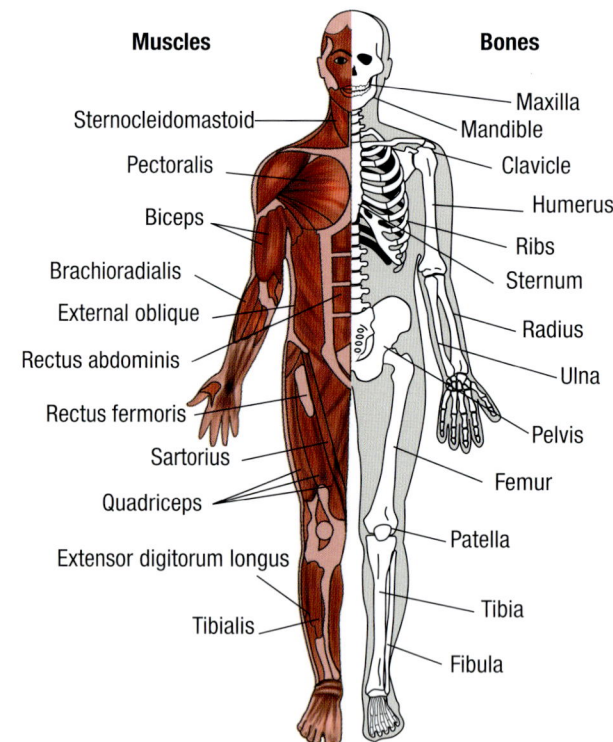

FIGURE 3.5 The human musculoskeletal system

Your **musculoskeletal system** provides form, support, stability and movement to your body. It is made up of the bones of the skeleton, muscles, cartilage, tendons, ligaments, joints and other tissue that holds your tissues and organs together. You are born with approximately 300 bones, although some fuse together as you get older.

On average, you have more than 600 muscles that make up about 40% of your body mass. Muscles cannot push – they pull. You can walk backwards and forwards or chew up and down because **muscles** often work in pairs called **antagonistic** (or opposing) **muscle pairs**. One muscle of the pair moves a body part in one direction and the other muscle of the pair moves it in the opposite direction. For example, the **biceps** muscle moves the forearm towards the upper arm by contracting fibres in the muscle cells. As the biceps muscle contracts (pulls), the **triceps** muscle relaxes to allow that movement to occur. To move the forearm away from the upper arm, the biceps muscle relaxes and the triceps muscle contracts. How the muscle moves depends on the way in which it is attached to the **skeleton**. Muscle is attached to bone by tough fibrous tissues called **tendons**.

Muscle pairs

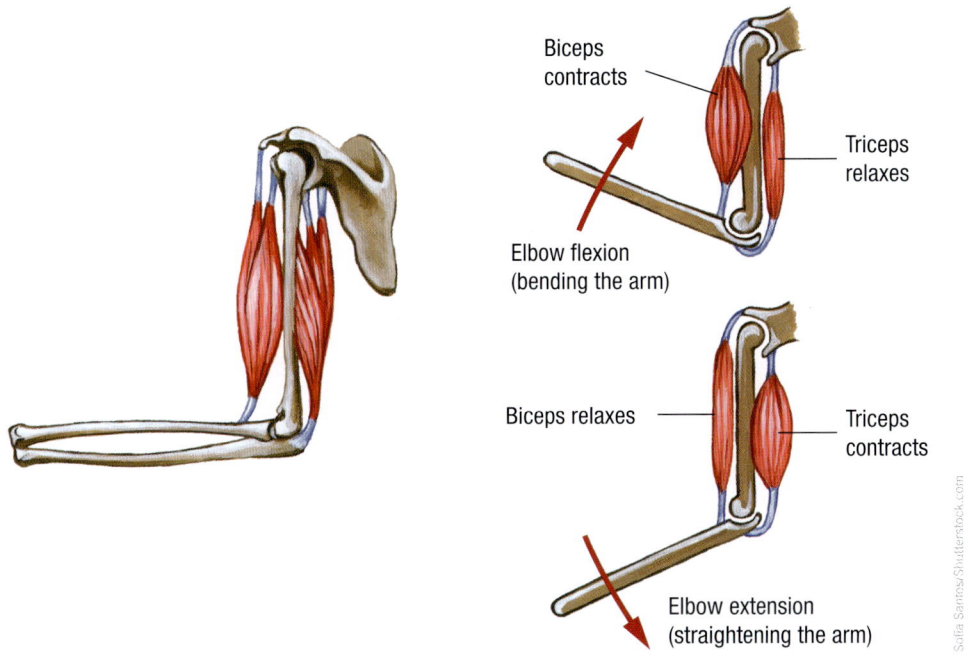

FIGURE 3.6 The triceps and biceps muscles work as a pair to bend and straighten the elbow.

Go to http://mypsci3.nelsonnet.com.au and click on **Interactive body**. Use this online simulation to locate and move muscles.

Muscles help you move, digest your food, pump blood around your body and breathe. Some of these actions you can control. For example, to jump or walk you use **skeletal muscles**, which are voluntary muscles. They are generally controlled by your conscious thoughts and decisions about movement and posture, though they can also be activated involuntarily through reflexes. They are connected to the skeleton, generally by tendons.

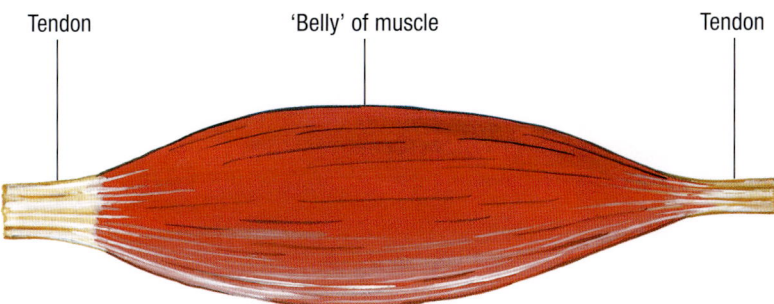

FIGURE 3.7 Skeletal muscles

FIGURE 3.8 Stretching muscles before and after exercising is important to prevent damage.

Other functions, such as breathing, are carried out by muscles over which you have no direct control. These are the **involuntary muscles**. **Smooth muscles** are involuntary muscles that are located within organs. It does not take thought to move these muscles. The movement of your diaphragm when you breathe and blinking your eyes are automatic. When you swallow, the involuntary muscles in your oesophagus use **peristalsis** to move your food down. Cardiac muscle is another type of involuntary muscle that is only located in the heart. Involuntary muscles make life much easier for you. Imagine if you had to remember to make your heart beat, your lungs inhale and exhale and your intestines digest food!

All the muscles in your body are made of bundles of muscle cells. They can contract (shorten) and stretch (lengthen). They do this when they receive a message from your brain. After a muscle has been stretched or contracted, it will go back to its original length, like an elastic band when you stretch it then let go.

Muscles can get bigger if they do work. Athletes spend a lot of time working out with weights and machines to increase the size of their muscles. If muscles are not used, they will waste (get smaller). The saying 'Use it or lose it' definitely applies to muscles.

TA HOW CAN IMPROVING YOUR HEALTH MAKE A DIFFERENCE TO YOUR COMMUNITY?

An increasingly popular way to support local efforts, charities and non-profit groups is by raising money or finding sponsors and competing in sports events. One such event, the World AIDS Marathon, has occurred each year since 2004. It seeks to raise not only money but also awareness about the AIDS epidemic. You could get sponsors and run in a similar event in your community or even organise a similar athletic event to raise money for a charity.

Diet for adolescents

It is important to eat a balanced diet with lots of vegetables and fruits. They provide the needed fibre in your diet and different vitamins and minerals you need to be healthy. Unfortunately,

FIGURE 3.9 Eating a variety of healthy foods is important to good health.

not everyone likes them or can afford them. Processed and packaged foods have become increasingly cheaper than fresh foods. Low-income areas may have only convenience stores and not grocery stores. Fast food offerings may be tasty, but they are loaded with sugar, carbohydrates and fat. Eating healthy meals on a regular basis takes commitment and planning. With busy schedules, it is easy to rely on fast, cheap food, but the long-term consequences are extra kilograms, high blood pressure and cholesterol, and more. The risk for adolescents is falling into bad habits such as skipping meals, excessive snacking and use of fast foods, and not following a balanced diet. Excessive sugar-sweetened drinks and other high-carbohydrate snacks are a particular risk.

ACTIVITY: Considering how healthy your diet is

In *Science 2 for the international student* Unit 3, you studied the digestive system and food nutrients. This activity asks you to consider your diet from the advice given in Figure 3.10 and the summary written below it.

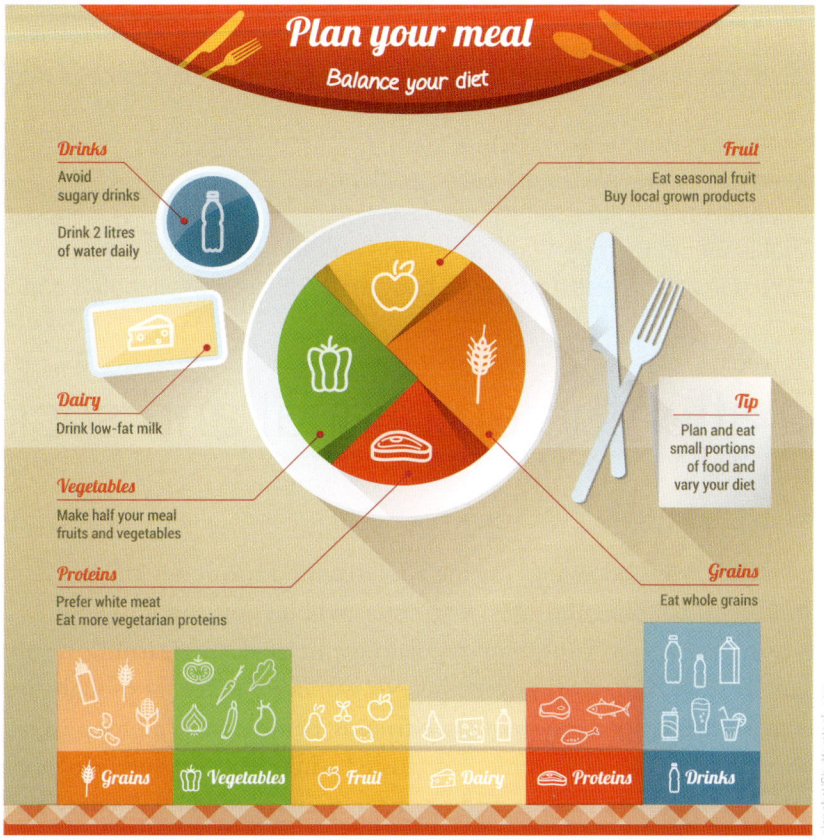

FIGURE 3.10 A summary of advice on diet to adolescents

Adolescents need a varied diet based on the five food groups listed here. The amount needed depends on body size and activity.

The daily diet of an adolescent should include close to 5–6 servings of vegetables; 3–4 servings of dairy products; 5–6 servings of bread, cereals and grains; 2 servings of fruit; and 2–3 servings of meat.

Adolescents need a lot of water. They should avoid soft drinks, fruit juices, sports drinks, energy drinks and tea or coffee.

1. Compare your diet with this advice.
2. Comment on the positive features of your diet.
3. Comment on the negative features of your diet.
4. Can you give any reasons why you think your diet is good, or is not good?
5. Has your diet changed over the past two or three years? Explain your answer.
6. Why do you think our diet is so important to our lives?
7. Would you like to improve your diet? In what ways could it be improved? How could you show the determination needed to improve it?

FIGURE 3.11 Regular exercise will make you feel better physically and emotionally.

> **REVIEW**
>
> 1. Outline the main features of adolescence and puberty.
> 2. Outline the function of the endocrine system.
> 3. List the organs of the endocrine system and state where each is located.
> 4. Describe what hormones are and how they affect the body.
> 5. State a function of the pituitary gland.
> 6. Outline what follicle-stimulating hormone (FSH) is and what its effects are in males and females.
> 7. Describe the features of a healthy diet.
> 8. List reasons why some people do not eat a healthy diet.
> 9. Outline the benefits of regular exercise.
> 10. Describe what is meant by muscle pairs. Include an example in your answer.
> 11. Describe some dangers of measuring your body image against a celebrity's appearance.

Looking after the emotional and social you

The emotional you

Go to http://mypsci3.nelsonnet.com.au and click on **Kids health** for tips on talking to your parents about things that matter. You'll also find lots of other helpful articles there.

You may experience crazy mood swings during adolescence. Hormones are mainly to blame. Growth hormones, sex hormones and stress hormones will be pouring into your body. In addition, your **limbic system**, a system in your brain regulating emotions and learning, is still developing. The reward system in your brain works overtime. No wonder you feel as though you are on an emotional roller coaster! Having a group of close friends, being part of a group in or out of school, and having an open relationship with parents or trusted adults can be a big help during this time. It is important to talk about your feelings and not keep them bottled up. Anxiety and depression in adolescence can easily lead to long-term mental disorders, and to substance abuse.

FIGURE 3.12 Playing sport increases your physical, emotional and social wellbeing.

The social you

Many schools offer a variety of extra activities, or in some cultures it is common for families to seek out after-school activities. Athletics, swimming, other sports, technology, scouts, band, choir, dance, art and theatre are common classes or activities. Schools may sponsor evening get-togethers such as dances. Sometimes, it's hard to find the balance between your interests and the amount of time you have available to devote to them. Sometimes, making good decisions in new social situations can be challenging.

Balancing the increasing rigour of school and the rewards of clubs can be tricky. You may find the close friends you had in the past have different interests and join different groups. It is normal to find your group of close friends changing during this time. Research shows many benefits from joining groups, including increased self-esteem, higher grades, more developed leadership skills, and a decreased likelihood of using drugs and alcohol. It is important to choose groups that will have a positive influence on you and help you make good decisions.

Peer pressure

During adolescence, **peer pressure**, the influence from others to change your attitude or behaviour, can be more prevalent than before. At times it can be good, giving you the courage to try new things that are okay – a new book or attractive hairstyle. But sometimes peer pressure can be negative, encouraging risky behaviour, such as using alcohol and drugs, and smoking. Saying 'No' may make you feel uncomfortable or worry that you might not be included in the group. It is important in these situations to follow what you know is right. You will have to live with the consequences; many of these behaviours can alter your life path permanently. Choosing groups that will support good choices can help prevent uncomfortable situations.

Go to http://mypsci3.nelsonnet.com.au and click on **Peer pressure** to learn more. Read the tips and take the online quiz 'Grind your mind' about clichés, making good choices and being true to yourself.

FIGURE 3.13 Being part of a social group with similar interests can have positive health benefits.

> **REVIEW**
>
> 1 Outline how the limbic system affects the adolescent body and the repercussions these effects might have.
> 2 Outline why it is important to talk about issues with others you trust.
> 3 Describe some benefits of being part of a club or social group.
> 4 Outline the positive and negative aspects of peer group pressure.
> 5 Imagine you had a classmate who clearly was being badly influenced by a group of new friends. How would you advise this person about how to respond to this peer pressure?

The brain and adolescence

Pressure from a group or individual is hard to resist for most teenagers. Most of the bad decisions teens make are due to **peer** (or group) **pressure**. In addition, just like the limbic system, the **prefrontal cortex** of your brain is undergoing massive remodelling. And the two systems are not 'talking' fluently to each other yet! The frontal lobes in the prefrontal cortex are responsible for decision-making, reasoning, impulse control, working memory and advanced planning, among other functions. So, at a time when you are beginning to have to make a lot of important decisions, your brain is not really helping you. This is one of the reasons adolescents make what adults think are dumb choices and exhibit risky behaviour. Understanding the underlying reasons may help you slow down and try to make better decisions.

Adolescence is the only time since birth that your brain is actually growing **neurons** and **synapses**. Neurons are the specialised brain cells that send electrical messages. Synapses are junctions between neurons: the pathways the electrical messages take. Over the course of your adolescence, your brain will prune out a lot of them, particularly in the frontal lobes. The old saying 'use it or lose it' is particularly true during this time. Only the pathways that get used get strengthened. The unused ones get pruned out. This may be why learning a new language during this time is much easier than after adolescence. This system will not be fully mature until you are in your 20s, making your teenage

Go to http://mypsci3.nelsonnet.com.au and click on **Reward pathway** to see how this works in your brain.

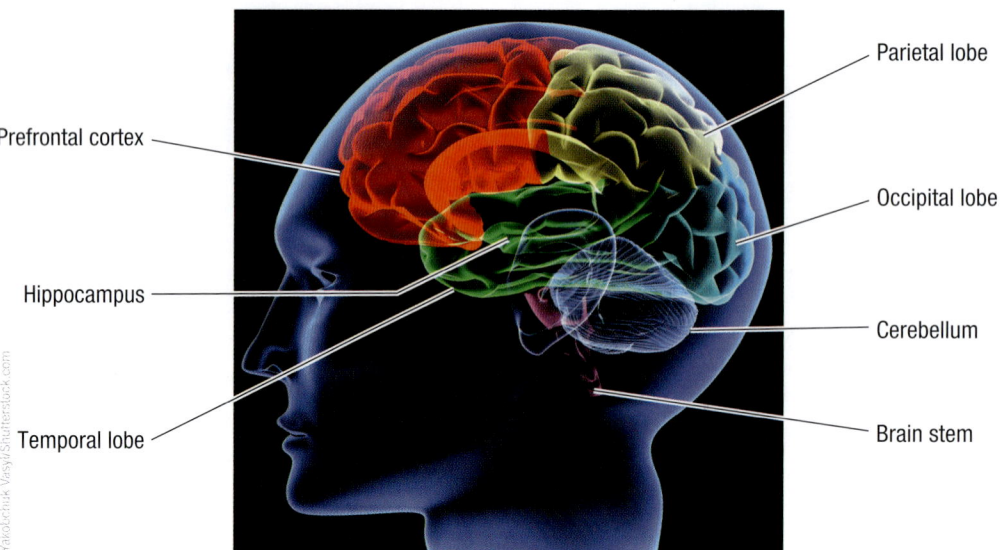

FIGURE 3.14 Major areas of the brain

years ripe for making mistakes! During adolescence, your brain is also extremely sensitive to the effects of drugs and alcohol. Research shows that addictions generally begin during adolescence and worsen during adulthood. For that reason, it is wise to steer clear of situations where you may feel pressured to use drugs and alcohol.

Drug misuse and abuse

Prescription drugs

When you feel sick, your parent or a doctor may give you medicine. This could include antibiotics, painkillers such as paracetamol or antihistamines for hay fever. These **prescription drugs** are designed to heal your specific illness or make you feel more comfortable, such as after surgery. Just as with alcohol, some people misuse drugs and may offer them to you or your friends using peer pressure. Taking drugs that are not prescribed by your physician or given to you by a parent can be very dangerous, especially to an adolescent body, which is going through changes at a rapid rate. Prescribed drugs, as well as over-the-counter drugs, should only be taken as directed. You should never take prescription drugs belonging to someone else.

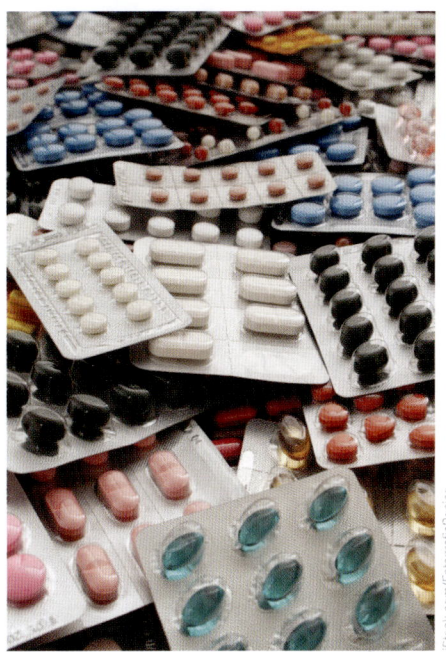

FIGURE 3.15 Prescription drugs

Addiction and tolerance to drugs

When a person continues using a drug, their body becomes accustomed to it and the effects become less powerful. Larger doses are then needed to achieve the same effects. This is called **tolerance**. Continued use can lead to dependence and, when stopped, withdrawal symptoms. This is called addiction.

Opiates and stimulants

Illegal drugs such as cocaine, morphine and heroin are illegal for a reason. Morphine and heroin are called opiates as they originally came from the opium poppy. Morphine is an important painkiller. Both morphine and heroin are extremely addictive and can easily lead to tolerance. This can in turn lead to an overdose. They are often injected, which can lead to people sharing needles and risking the transfer of diseases such has HIV and hepatitis B.

Cocaine is made from the leaves from the coca plant, which grows in South America. Cocaine is a very strong stimulant that provides people with a short-lived but intense high. It is an extremely dangerous drug. It is extremely addictive and can lead to nervousness, lack of sleep and paranoia, and long-term use can lead to high blood pressure; heart attacks; damage to liver, kidneys and lungs; damage to the nose (if inhaled through the nose); hallucinations; sexual problems and severe depression.

Go to http://mypsci3.nelsonnet.com.au and click on **Addiction** to explore 'The Science of Addiction: Genetics and the Brain'. You will find a number of interactive simulations and interesting articles on the adolescent brain.

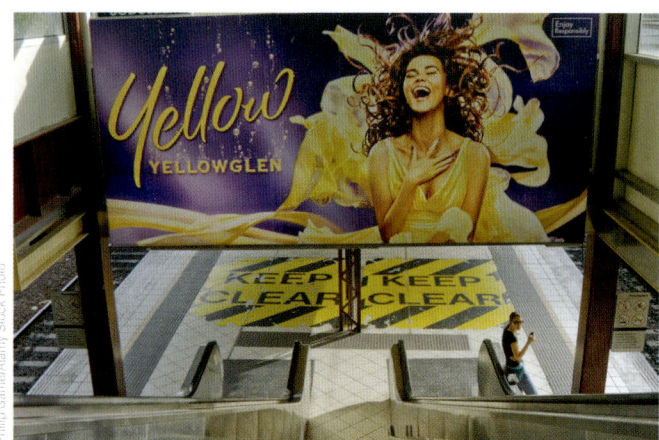

FIGURE 3.16 An advertisement for alcohol – what is the message?

Alcohol – one of the greatest health challenges in our lives

You cannot open a magazine or watch a sports broadcast without being bombarded with commercials for alcohol. Many adults drink alcohol, so why not adolescents? One reason is that the adolescent brain processes the effects of alcohol much differently from an adult brain. But, as is explained below, a high percentage of adults have had their lives negatively affected by alcohol. Although alcohol is a 'legal' drug, it has enormous potential to be misused or abused.

People drink alcohol to be more relaxed and less inhibited. This can make shy people feel more sociable, although this increased sociability is likely to be out of character. As people drink more, their judgement is often affected and they may feel uncoordinated and have trouble seeing or hearing. Their bodies struggle to cope, they slur their speech, and they can have trouble standing and walking. They can become aggressive and do or say something they regret later. For instance, they may have unwanted and unprotected sex with serious consequences such as unwanted pregnancy. Hangovers are caused by the dehydrating effect of alcohol on the brain and body.

Some of the other social and mental effects of using alcohol in adolescence include decreased learning potential, memory and hormone production and increased risk of depression and social problems. It leads to adults missing time at work and generally not being able to cope with work or family life. Excessive alcohol use is also associated with diseases of the liver, kidneys and heart, and increased risk of cancer, including throat, stomach, colon, breast and kidney cancer. If pregnant women consume alcohol, it can damage the foetus and lead to birth defects. Alcohol is enormously addictive and millions of people worldwide are seriously addicted to alcohol; that is, they are alcoholics.

The costs to society of excessive alcohol use are enormous. If alcohol was invented tomorrow, it probably would not be given permission to be used.

The use of alcohol is very culturally linked. In some religions, its use is not tolerated. In some societies, you rarely see drunk people in public; in others, public drunkenness is very common.

FIGURE 3.17 People with alcoholism are unable to control their drinking.

Some people in Asia have a gene that causes their body to metabolise alcohol in a different way from other people, which encourages them not to like alcohol. It seems that there is also a gene that makes people more prone to **alcoholism**, so alcoholism runs in families. Countries have different age limits for legal drinking, or the buying of alcohol. Many do not allow it until you are 18 or 21.

Binge drinking (drinking a large amount of alcohol with the purpose of getting intoxicated) is a known problem among adolescents. This is particularly dangerous as it can cause serious medical harm, and also increases the probability of developing alcoholism as an adult. Every year, young people die from binge drinking. Unfortunately, the use of alcohol can be seen as a rite of passage, or as a way to fit into a group. Coping with peer pressure is a crucial challenge for adolescents in relation to alcohol.

> **ACTIVITY** **Helping adolescents make decisions about alcohol**
>
> Imagine the following situation.
>   One of your best friends is regularly drinking alcohol. Think about how you would advise him or her.
>   Write a role play describing the conversation you could have with him or her.

FIGURE 3.18 Common substances such as coffee, alcohol and cigarettes are classified as drugs that can be dangerous when abused.

Cannabis – a particular risk for adolescents

Cannabis comes from the cannabis plant. The plant has been used for over a thousand years, initially in Asia and the Middle East, but now all around the world, to make hemp. Hemp is used to make fibre for ropes, cloth and paper; healthy oils and bird seed. The variety of cannabis used to make hemp has almost no active ingredient and is a normal agricultural crop.

Some varieties of cannabis (also called marijuana, hashish, skunk, pot, a joint etc.) have an active ingredient called tetrahydrocannabinol (THC). This type of cannabis is a drug that has been used for recreational purposes for some time, particularly since the 1960s. In recent times it has also been used for medical purposes, such as the treatment of multiple sclerosis and cancer. It has the effect of making people feel happy and relaxed. Also it affects people's senses, and people claim it increases their enjoyment of music and food. There has always been a lot of debate about how dangerous it is and whether it should be legalised.

However, for adolescents cannabis has special risks. As described earlier, the adolescent brain is a brain in construction. The use of cannabis can have some real, harmful effects on an adolescent brain. It can make people feel very anxious and sometimes paranoid. Regular use makes it very difficult to concentrate and it affects your memory. Regular users of cannabis can find it very difficult to succeed with their studies at school. It also seems to affect people's motivation and regular users can become apathetic.

Like tobacco, cannabis is linked to lung disease, particularly lung cancer. People often smoke cannabis with tobacco, which further increases the health risks. Its use has also been associated with lack of fertility in men, and increased blood pressure.

People might tell you cannabis is natural and safer than other legal drugs such as cigarettes. Be careful – there are many myths about cannabis. Recent research is showing that its use in adolescence is causing more problems and mental illnesses than we realised. And, of course, it is addictive.

Go to http://mypsci3.nelsonnet.com.au and click on **Drug effects** to play the interactive Mouse Party looking at the effects of drugs on mice.

FIGURE 3.19 Cannabis plants being grown for hemp

FIGURE 3.20 Alcohol and cigarettes – not a way to good health!

REVIEW

1. Summarise the effects of alcohol on:
   a the body
   b society.
2. Describe why the prefrontal cortex is such an important part of the brain, and why its development is so important during adolescence.
3. When can prescribed drugs be dangerous?
4. Describe the reasons why adolescents should be very concerned about the possible effects of cannabis.
5. Why are morphine, heroin and cocaine considered particularly dangerous drugs?
6. What effect does substance abuse have on your ability to do well at school?

Smoking tobacco

The following are some facts about smoking tobacco.
- Tobacco kills up to half of its users.
- Tobacco kills around 6 million people each year. More than 5 million of those deaths are the result of direct tobacco use while more than 600 000 are the result of non-smokers being exposed to secondhand smoke.
- In most countries, tobacco causes more deaths every year than HIV, illegal drugs, alcohol, motor vehicle accidents and firearm accidents combined.

- Smoking increases the chance of lung cancer by 25 times.
- Smoking increases the chance of heart disease by 2–4 times.
- Smoking can cause cancers almost anywhere in the body.
- Smoking will seriously affect a baby's health before and after birth.
- Smoking affects sperm, teeth and gums, and increases your chance of developing diabetes, arthritis, chronic bronchitis, emphysema and pneumonia.
- Nearly 90% of all smokers started smoking in their teenage years.
- Smoking is very expensive; a pack a week will cost you over $700 a year, a pack a day over $5000 a year.

Tobacco smoke contains nicotine, a powerful drug that affects several organs, including the brain, heart and lungs. Nicotine activates the area of your brain dealing with pleasure and reward. It releases a chemical called **dopamine**, which is called the 'pleasure molecule'. This is one reason why people get 'hooked' on smoking. They become addicted to tobacco just as they can with other drugs. Cigarette smoke also contains tar, carbon monoxide and other chemicals that can limit the amount of oxygen carried in the blood. The tar contains chemicals that cause cancer and also affect lung function. Smoking doesn't only affect the person doing it. Secondhand smoke from being around someone who smokes can be just as dangerous over a long period.

The story of how scientists were able to demonstrate the health effects of smoking is very interesting. It is not easy to isolate one factor, such as smoking, when studying common diseases such as lung cancer and heart disease, which can be caused by many factors. In science, this is called establishing 'cause and effect'. Scientists need to show that the more you smoke the higher the chance of developing lung cancer or heart disease, but also they need to demonstrate a scientific reason for smoking producing these diseases.

Effect of smoking tobacco on the lungs

Tobacco smoke can damage the lungs in a number of ways. To help you understand these effects, we will study the respiratory system.

How the lungs work

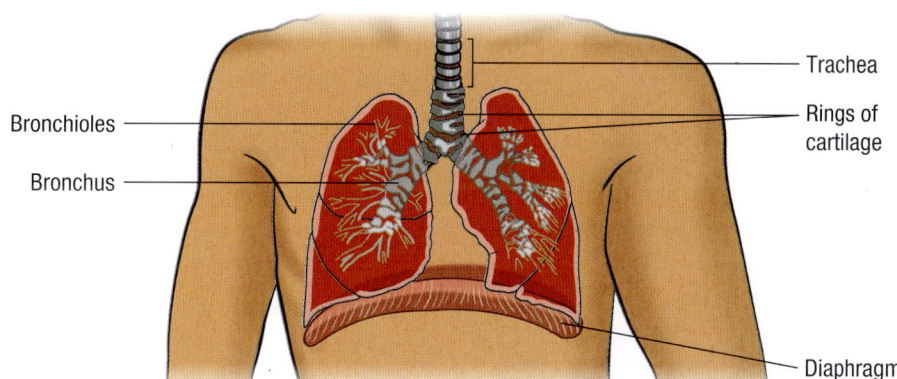

FIGURE 3.21 The lungs

All the cells in your body need a constant supply of oxygen. It is the role of the **respiratory system**, working together with the **cardiovascular system**, to supply oxygen to the cells and to dispose of carbon dioxide as waste.

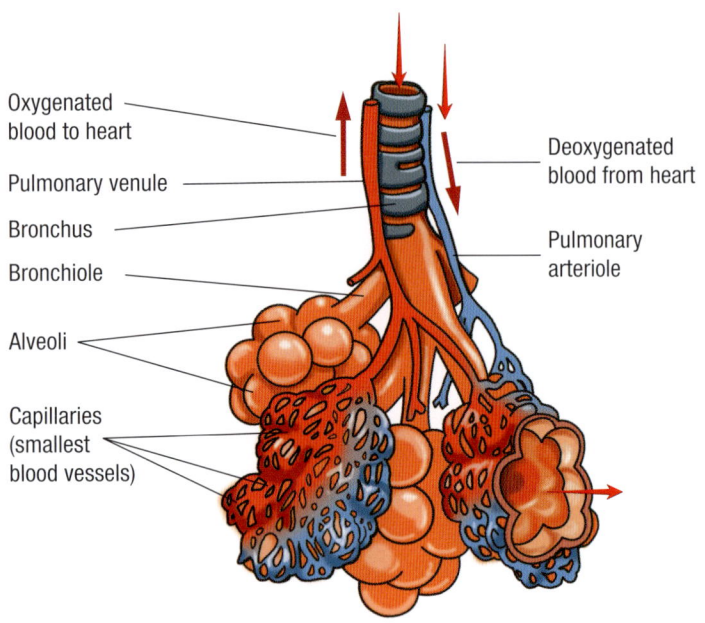

FIGURE 3.22 Alveoli

The hero of the respiratory system is the **lung** and its unique structure. The lungs are paired organs, each 25–30 centimetres in length. The windpipe, also known as the **trachea**, divides into two major **bronchi** (singular **bronchus**), one of which goes to each lung. The bronchi then divide and divide again inside the lung. Your respiratory system consists of a system of tubes that get narrower and narrower. At the ends of these narrow tubes (the bronchioles) are your **respiratory surfaces**, the surface membranes of millions of tiny air sacs, known as **alveoli** (singular alveolus). Each alveolus is only about 0.2 millimetres wide. If you were to lay out the surface area of all the alveoli within your lungs, they would cover approximately 65–70 square metres – an area roughly a third of the size of a tennis court. Blood pumped through this moist environment is easily oxygenated and waste carbon dioxide is released.

The upper part of the respiratory system (trachea and bronchi) contains cilia. Cilia are tiny hairlike structures that move in a swaying motion to trap and remove foreign debris and microbes. The cilia secrete a sticky mucus that traps all these harmful particles and then moves the mucus out of the lungs. Tobacco smoke can damage the cells and these protective **cilia**. This then causes the mucus to build up in the lungs, which causes the typical smoker's cough and can lead to lung infections.

Carbon monoxide in the cigarette smoke reduces oxygen levels in the blood. This means your cells receive less oxygen for respiration so you are likely to get breathless when you exercise. This also causes developing babies in the womb to receive less oxygen, which is particularly dangerous.

Cigarette smoke contains many different chemicals that irritate the lining of the bronchi and bronchioles. The lining becomes inflamed and can become infected. This causes chronic bronchitis. Tobacco smoke contains tar, a black sticky substance. This coats the lining of the alveoli, which results in less surface area for gaseous exchange. This makes you breathless, especially after exercise. It also leads to emphysema, for which some people eventually need a special oxygen supply. Some of the chemicals in tobacco smoke are **carcinogenic** (cause cancer). Smoking causes 90% of all lung cancers. The cancers grow slowly and often there are no symptoms until they have spread to other parts of the body, forming more serious secondary tumours.

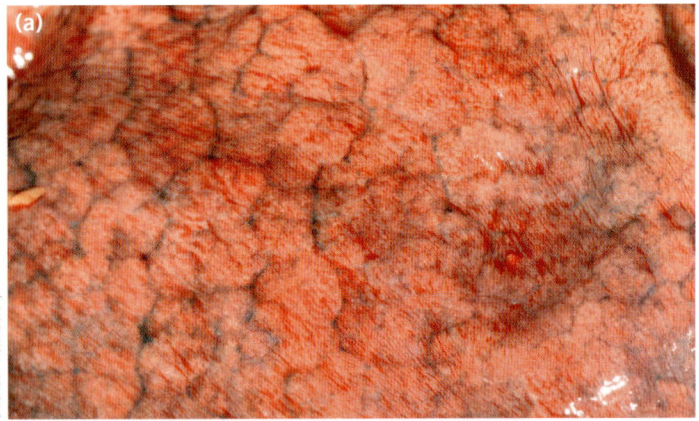

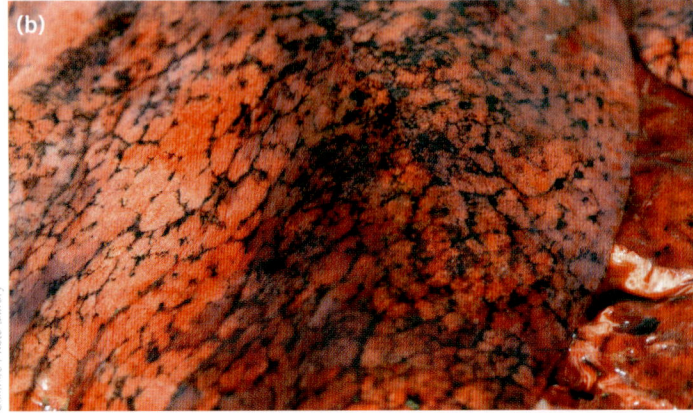

FIGURE 3.23 (a) A normal lung and (b) a smoke-damaged lung

How air gets in

Holding your breath is at first easy, but it becomes more difficult the longer you hold it. This is because as your cells break down sugar to release energy, carbon dioxide waste is produced and accumulates in your blood. Carbon dioxide is removed in the lungs. As carbon dioxide accumulates, this causes you to breathe out (exhale). Your body also requires oxygen in order to release the energy for your cells to do work. It is essential that oxygen is taken from the air.

Take a deep breath. As you breathe in (inhale), your **diaphragm** contracts and moves downward, allowing your lungs to expand (Figure 3.24). It is impossible to breathe and not have your diaphragm rise and fall. Try it.

After each breath in, you must also breathe out. Exhaling and inhaling occurs every few seconds, or faster if you are exercising.

How you manage to breathe air in and out is the result of a coordinated effort between the muscles that link one rib to the next, known as the intercostal muscles, and your diaphragm, which is a sheet of muscle.

Put your hand on your chest and breathe in. Can you feel your lungs get larger?

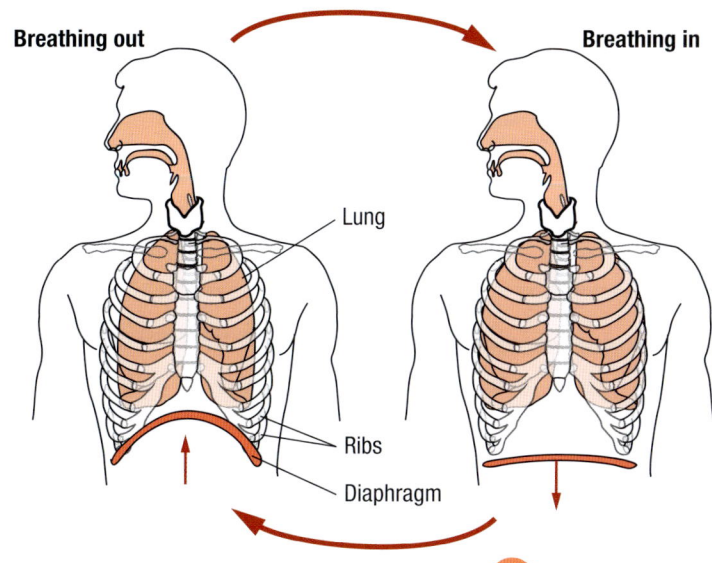

FIGURE 3.24 Inhaling and exhaling

CRITICAL THINKING
Reflectively analysing evidence: use these graphs to discuss the idea of 'cause and effect' in science.

ACTIVITY: Graphs relating to smoking and health

TASK
Examine the three graphs in Figure 3.25 relating to smoking. What claims can be made from the evidence presented in each of these graphs? What further questions would you have of each study?

Go to http://mypsci3.nelsonnet.com.au and click on **Smoking** to play a game of Smokes and Ladders to find out how much you know about the dangers of smoking tobacco.

Go to http://mypsci3.nelsonnet.com.au and click on **Dopamine** to find out how your brain reinforces behaviours by releasing dopamine.

Go to http://mypsci3.nelsonnet.com.au and click on **Quit smoking** for advice to adolescents on giving up smoking.

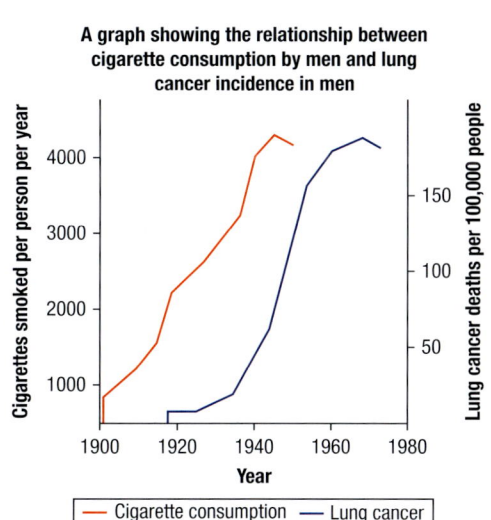
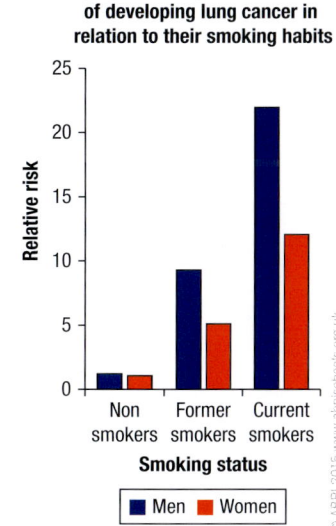
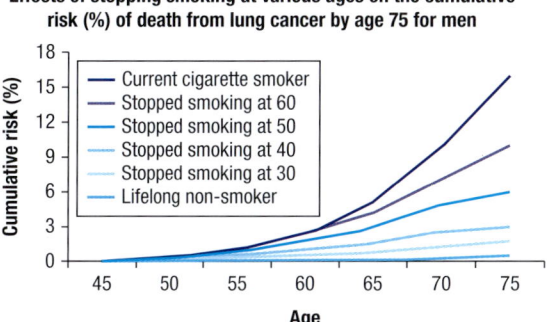

FIGURE 3.25

Go to http://mypsci3.nelsonnet.com.au and click on **Environmental tobacco smoking experiment**.

Smoking experiment

ACTIVITY

TASK
Visit the weblink to perform an activity in which you will collect and analyse scientific data from an experiment in which mouse foetuses are exposed to environmental tobacco smoke. You will use the steps of a scientific experiment to collect data, and interpret your data.

What factors affect lung capacity?

INVESTIGATION 3.1

YOUR CHALLENGE
Design an investigation that tests what variables affect the lung capacity of students.

THIS MIGHT HELP
Your teacher will provide you with a lung volume bag with a mouthpiece and mouthpiece holder to aid in collecting data (a balloon can work just as well).

THINGS TO THINK ABOUT
- How can you measure the volume of gas exhaled?
- How will you collect the air?
- How will you determine its volume?
- What are some variables you could test?
- Does exercising or breathing cold or moist air change your lung capacity in the short term?
- Does your size or gender affect your lung capacity?

Carry out and write up your investigation following the guide in Appendix 3 on page 177 or as advised by your teacher.

SAFETY
- Be mindful of sterilising mouthpieces between users.
- Let your teacher or nurse know if you have **asthma**.

REVIEW

1. List the ingredients of tobacco smoke that are harmful to your health.
2. Describe how people can become addicted to cigarette smoking.
3. Outline how people around a smoker can be harmed, even if they do not smoke.
4. Outline details of how the cilia in your lungs are affected by cigarette smoke and the health consequences.
5. Describe the function of the lungs, including diagrams.

UNIT QUESTIONS

CRITERION A

EXPLAINING SCIENTIFIC KNOWLEDGE

1. State five ways in which tobacco smoke can damage the body over time. (Level 1–2)
2. Identify each from the clue. (Level 3–4)
   a. Cells in the brain that send electrical signals
   b. The developmental stage that is generally your second decade of life
   c. Area of the brain responsible for impulse control and decision-making
   d. Tiny hairs in the respiratory tract that move back and forth
   e. The developmental stage in which your body sexually matures
   f. One of the systems of the brain responsible for emotion and learning
   g. Pathways in the brain that carry electrical signals from one cell to another
   h. The influence from others your age to change your attitude or behaviour
   i. The chemical called the 'pleasure molecule'
3. Outline how lungs inhale and exhale air. (Level 5–6)
4. Outline why emotional and social health is very important during adolescence. (Level 5–6)
5. Explain why adolescents are more susceptible to the effects of substance abuse. (Level 7–8)
6. Describe in detail the function of muscles. (Level 7–8)

APPLYING SCIENTIFIC KNOWLEDGE AND UNDERSTANDING TO SOLVE A PROBLEM

7. The air you breathe in has a different composition from the air you breathe out. Table 3.1 shows the percentage of oxygen, carbon dioxide and water vapour in your exhaled breath.

   TABLE 3.1 What is in our breath?

Gas	Percentage in inhaled air	Percentage in exhaled air
Oxygen	21	16
Carbon dioxide	0.04	4
Water vapour	Variable	Saturated (cannot hold any more)

   a. State the difference in the percentage of oxygen in inhaled and exhaled air. (Level 1–2)
   b. State the difference in the percentage of carbon dioxide in inhaled and exhaled air. (Level 1–2)
   c. Suggest a possible reason for the difference in percentage of oxygen in inhaled and exhaled air. (Level 3–4)
   d. Suggest a possible reason for the difference in percentage of carbon dioxide in inhaled and exhaled air. (Level 3–4)

8. Explain why it is advantageous to be part of a club or social group, such as athletics, band or a maths club. (Level 3–4)
9. You feel as though you are getting bronchitis, a serious inflammation and infection of the lungs. A friend offers you medicine she has left over from when she was recently sick. Discuss why it would be a good or bad decision to take this medicine. (Level 5–6)
10. The incidence of substance abuse (alcohol and drugs) is increasing. How can you re-educate members of your school and community so that students feel comfortable saying, 'No, that's not for me'? (Level 7–8)

INTERPRETING INFORMATION

11. Your parents think that peer pressure is a negative force. How can you convince them otherwise? (Level 1–2)
12. Friends tell you that binge drinking only a few times won't hurt you. Outline the science on whether or not this is true. (Level 3–4)
13. You and your parents are disagreeing about some of your decisions. Describe how you can explain the emotional and physiological changes you are undergoing to help them support you to make better choices. Use recent research on the adolescent brain. (Level 7–8)

14 Figure 3.26 shows some data on how lung volume of men and women changes with age. (Level 5–6)

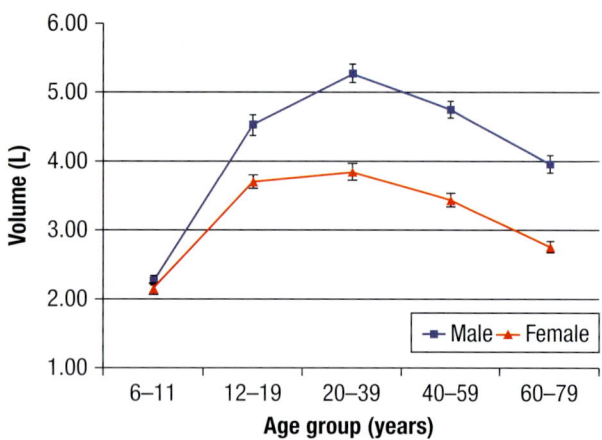

FIGURE 3.26 Maximum volume of gas that can be exhaled from a full inhalation

a Describe what this data shows about how lung volume changes with increasing age.
b Describe what this data shows about how lung volume differs between men and women.
c Give scientific reasons for these conclusions.

REFLECTION

1 In this unit, you have studied the possible consequences of personal decisions relating to diet, exercise and drug use during adolescence. In what other areas are there consequences from the way we live our lives?
2 In this unit, you have seen models to show the respiratory and musculosketal body systems. Your teacher may have used physical models. What advantage does the use of models have in education?
3 Discuss how the concepts of form and function can be applied to the function of the lung.
4 What are your thoughts about the suggestion that the adolescent years are the most challenging of our lives?
5 Explain why the concept of cause and effect is very important in the sciences. Give some examples.

UNIT 4

USING METALS

KEY CONCEPT
Relationships

RELATED CONCEPTS
Form

Function

Consequences

GLOBAL CONTEXT
Scientific and technical innovation – an exploration into the properties of metals and how these affect the use of metals

STATEMENT OF INQUIRY
Scientific and technical innovations utilise specific properties of metals in order to ensure the desired outcomes.

INQUIRY QUESTIONS

FACTUAL
1. What are metals, non-metals and metalloids?
2. What are the physical and chemical properties of metals?
3. What are the different types of metals?

CONCEPTUAL
4. How do the properties of metals affect their use?
5. How do we explain that metals have different properties?

DEBATABLE
6. Why do we have a fascination with gold?
7. Given modern scientific knowledge, why is rusting still such a problem?

Introduction

Consider how your life would be different if there were no wires to conduct electricity. Or if the saucepan melted when you boiled water. Or if all fireworks were the same colour. What if you could not walk because you couldn't get a knee replacement? Or if your gold ring dissolved when you washed your hands? These examples show a few of the uses of metals, and how the properties of the metal make it suitable for that use. Metals play many important roles in our lives, including being vital for our health and making many medical procedures possible. In this unit, you will look at what metals are, their properties and how they are utilised.

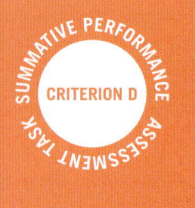

CRITERION D
SUMMATIVE PERFORMANCE ASSESSMENT TASK

ATL

INFORMATION LITERACY: REFERENCING

Researching is a very important skill to develop both as a scientist and as a student. When you research, you need to use a range of resources so that you get more than one point of view. It is important to acknowledge where the information you use has come from. This is why we use bibliographies and in-text citations.

Metals in our body

As our understanding of different metals and alloys has developed, so has our use of them. This is because we can match a metal to a use on the basis of its properties. Nowhere is this more important than in our bodies. Prosthetic limbs, pins to fix a broken leg and replacement joints may all be made out of metal. In these situations, the metal must be able to perform its function effectively and safely.

Choose one situation in which metals are used in our bodies. Research what metal is used, how it is used and why it is suitable for this use. Consider what factors (moral, cultural, ethical, political, environmental and/or economical) influence its use and how this may vary in different situations. Negotiate with your teacher as to how you will present your findings in an effective manner. Make sure that you correctly reference your sources using both in-text referencing and a bibliography.

Metals

As you learnt in *Science 2 for the international student*, different **elements** have different properties. However, some elements can be grouped together on the basis of their similar properties. One method of classifying elements is as **metals**, **non-metals** or **metalloids**.

This unit will look at the key ideas about metals – what they are, and what it is that makes a metal a metal.

Metals on the periodic table

Metal elements are found on the left-hand side of the periodic table and non-metal elements are on the right-hand side. As there are many more metal elements than non-metal elements, the division is not in the middle. Rather, it is a zigzagged line that moves downward toward the right side of the table.

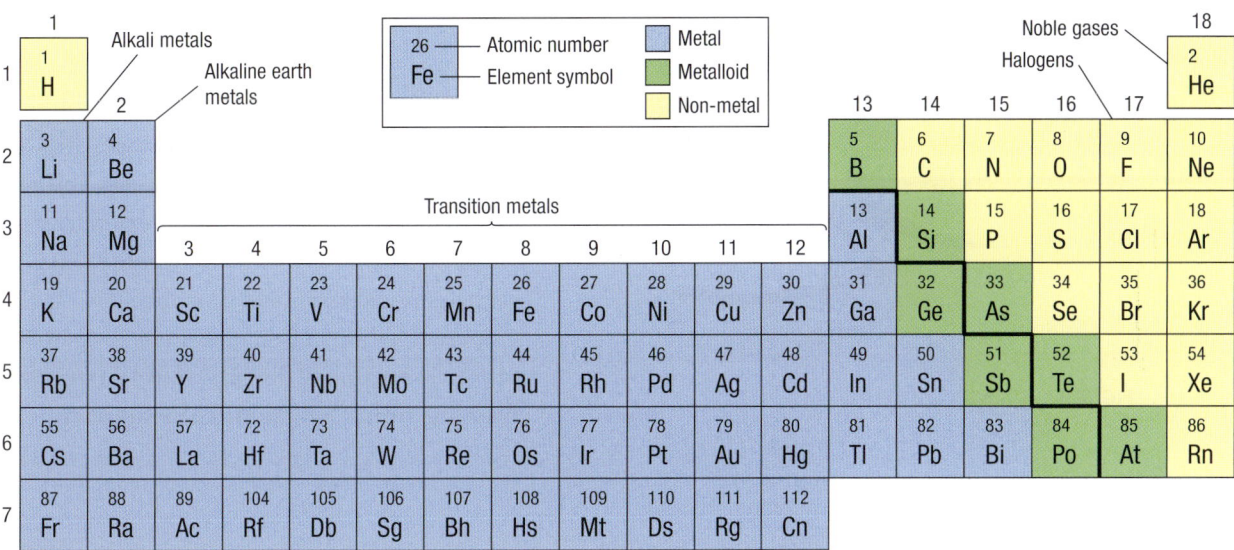

FIGURE 4.1 Part of the periodic table showing the location of metals, non-metals and metalloids

Properties of metals

You should remember from *Science 2 for the international student* Unit 4 that the elements in the periodic table are arranged according to their properties. Therefore, it makes sense that the metals share some properties. These properties are shown in Table 4.1.

TABLE 4.1 The properties of metals

Property	Meaning
Opaque	You cannot see through them.
Lustrous	They are shiny, as they reflect light.
Good **conductors of electricity** when they are a solid or a liquid	Electricity can pass from one end to the other.
Good **conductors of heat**	Heat can pass from one end to the other.
Malleable	They can be hammered into sheets.
Ductile	They can be drawn into wire.

FIGURE 4.2 The properties of metals make them useful for many different purposes. They (a) conduct electricity and (b) conduct heat.

FIGURE 4.3 Gold is a soft metal.

Properties of metals that vary

Even though metals share many properties, there are still variations between them. This is important when choosing metals for different purposes. However, metals in the same column of the periodic table have more similar properties. Table 4.2 describes some properties of metals: **hardness**, **density**, **tensile strength**, **melting point** and **chemical reactivity**.

TABLE 4.2 Properties that vary from metal to metal

Property	Meaning	Variations
Hardness	Resistance to being scratched The hardest substance cannot be scratched by any other material.	Most metals are fairly hard to scratch. Gold, lead and calcium are soft metals, i.e. they are easily scratched.
Density	The number of grams of the material in each cubic centimetre. If a material is denser than water, it will usually sink in water.	Density varies from metal to metal. For example, 1 cubic centimetre of gold weighs 19.3 grams, but the same volume of aluminium weighs only 2.7 grams. Gold is a very dense metal, whereas aluminium is a 'light' (not very dense) metal.
Tensile strength	The ability of a material to withstand a stretching force For example, the materials in cables that are suspended between poles must have high tensile strength.	Tensile strength varies considerably from metal to metal. For example, steel (which is mostly iron) has a higher tensile strength than aluminium.
Melting point (MP)	The temperature at which the material melts, i.e. turns from a solid to a liquid	The MP varies considerably from metal to metal. Mercury is already a liquid at room temperature (its MP is −39°C) but a large number of metals require temperatures well over 1000°C to melt them.
Chemical reactivity	The ability of a substance to react with other substances	In the case of a pure metal, which is an element, any reactions it undergoes will produce compounds of the metal. A very small number of metals such as gold and platinum are very unreactive. Others such as silver, copper and tin are fairly unreactive. But most will react with water, some very slowly, like iron, and others explosively.

ACTIVITY — Metals in our daily lives

Metals are used all around you every day. Gold is often used for making jewellery as it is easy to mould, unreactive and an attractive colour. Copper is used in electrical wiring as it is a good conductor of electricity and ductile.

In groups of three or four, brainstorm all the items in your houses that you can think of that are made of metal. For each, discuss, and possibly research, the type of metal it is made from and the properties it must exhibit to be suitable for that purpose. Make sure you collaborate in your group by sharing your ideas, listening to others and dividing the work. As a group, decide how you will present your information in a way that is clear and easy to understand, as well as attractive to look at so that people will want to read it.

TRANSFER: MEMORY
Part of learning is remembering new information so that we are able to recall the information and apply it to questions or problems. There is a range of different strategies for memorising things. However, all rely on making connections in our brain between previous information and the new information. The weblink provides some strategies for improving your memory.

Go to http://mypsci3.nelsonnet.com.au and click on **Memory strategies**.

Types of metals

Metals can be classified into different groups according to their properties. These groups correspond with their placement on the periodic table.

Alkali metals

The **alkali metals** are found in group 1 of the periodic table and include sodium (Na), potassium (K), lithium (Li) and caesium (Cs). These metals are less dense than many metals: lithium, sodium and potassium will actually float on water, as they are even less dense than water. They are silvery coloured and soft – you can cut them with a knife. These metals are not found as pure elements in nature because they are very reactive. This means that they will react with the surroundings and form a compound. For this reason, the sodium in your school chemical storeroom will be stored in oil to stop it reacting with the oxygen in the air. The group 1 elements lower on the periodic table are more reactive than those higher up. Sodium can be explosive when it is placed in water and reacts. Imagine what potassium will do!

FIGURE 4.4 Sodium is stored in oil to stop it reacting with oxygen.

FIGURE 4.5 Potassium reacts explosively with water.

Alkaline earth metals

The elements in group 2 of the periodic table are the **alkaline earth metals** – they are all found in the Earth's crust as compounds in rocks. Calcium (Ca) is the fifth-most common element in the Earth's crust; magnesium (Mg) is the eighth-most common. They are quite reactive, but not as reactive as the elements in group 1. The alkaline earth metals are all silver-white metals that melt at higher temperatures than the alkali metals.

Transition metals

The **transition metals** make up groups 3–12, or the middle block of the periodic table. These metals are harder and denser than those in groups 1 or 2, and have higher melting points. Some of these metals will produce coloured compounds (Figure 4.6). For example, copper (Cu) compounds are green or blue, cobalt (Co) compounds are red and nickel (Ni) compounds are green.

FIGURE 4.6 Some transition metals produce coloured compounds.

Metals are selected for a particular use based on their properties. Table 4.3 show the uses of some transition metals.

TABLE 4.3 Transition metals and their uses

Transition metal	Use	Property
Iron	Building materials, usually in the form of steel (an alloy of iron)	Strong
Titanium	Joint replacements	Unreactive, strong, smooth
Copper	Electrical cables	Good conductor of electricity
	Water pipes	Easily bent, does not react with water
Silver	Jewellery	Does not corrode with water or air, maintains a shiny surface
	Circuit boards	Very good electrical conductor
Gold	Jewellery	Does not corrode with water or air, maintains a shiny surface
	Connecting wires for electrical circuits	Good electrical conductor

ACTIVITY: Mining for chocolate

In mining, the desired metal is rarely found in its pure state. In most cases, the metal has reacted to form compounds. These compounds are found in the rocks that are mined from the ground. We call these rocks **ores**, which means rocks that contain minerals (naturally occurring solid chemical substances) with important elements, including metals. During the process of mining, these rocks are removed and treated to recover the pure metal.

YOUR CHALLENGE
Your teacher will provide you with different brands of chocolate chip cookies; these represent ores. The biscuit part is your waste rock and the chocolate chips are your valuable mineral. Your challenge is to:
- determine the percentage of valuable mineral in your supply of ore
- find the most cost-effective and environmentally friendly way of extracting your mineral.

THIS MIGHT HELP
In your group, brainstorm how you will meet your challenge and give your teacher a list of what you need and a short description or flowchart of what you plan to do.

THINGS TO THINK ABOUT
1. Consider the properties of the waste rock and mineral. How can these be used?
2. The less mineral you waste, the less labour, energy and materials your process requires. The less time you take, the more profit you will make. Which of these features will also make your process more environmentally friendly?
3. How can you best record what you do to prove how effective it is?

WHAT DID YOU DISCOVER?
1. What percentage of your ore was the valuable mineral?
2. How successful were you in concentrating your mineral?
3. How long did you take to complete your task?
4. What quantities of materials did you consume in the process?

WHAT DO YOU THINK?
1. What properties of the waste rock and minerals did you use to separate them?
2. Of all the methods used by the groups in the class, which method of separation worked best and why?
3. Describe the problems you encountered and discuss how they could be overcome.
4. How might your process be similar to or different from the process used to extract a metal from its ore?

Colours of transition metal compounds

EXPERIMENT 4.1

AIM
To observe the colour of compounds containing transition metals and non-transition metals as a solid and in a solution.

MATERIALS
- range of solid samples of compounds containing transition metals or non-transition metals
- spatula
- distilled water
- test tubes (two per sample)
- test-tube rack
- stirring rod
- labelled beakers for waste chemicals

SAFETY ADVICE
- Wear safety glasses and protective clothing at all times.
- Do not pour any of the chemicals down the sink. Follow your teacher's instructions regarding the safe disposal of the chemicals.

PROCEDURE
1. Label two test tubes for each compound with its name or formula.
2. Place a spatula of the relevant compound into each of the test tubes.
3. Add about $1\,cm^3$ of distilled water to one test tube for each compound. Stir the mixture until the solid has dissolved.
4. Observe the colour of the solid and its solution for each compound, and record your findings in a table like the one shown in the Results section.
5. Place your chemicals into relevant waste beakers and clean up your bench.

RESULTS

Compound	Metal in compound	Type of metal	Colour of solid	Colour of solution

DISCUSSION
1. What did you notice about the colours of the compounds containing non-transition metals?
2. What did you notice about the colours of the compounds containing transition metals?
3. What did you notice about the colours of the solid and the solution of each compound?
4. Research why we see different colours. How might this apply to the colour of compounds?
5. How can knowledge of the colours of compounds be useful in science or our everyday lives?

CONCLUSION
What conclusion can you make regarding the colour of compounds of transition metals?

EXTENSION
Research how transition metals are used in fireworks.

Alloys

When a metal is mixed with other elements, it forms an **alloy**. Bronze, brass and stainless steel are three common alloys. The properties of alloys differ from those of the original metals. This allows alloys to be used for different purposes. For example, stainless steel is much more durable than pure iron, and so is used extensively in the construction of buildings. Gold jewellery is made of an alloy of gold. The term 'carat' refers to the purity of the gold: 24-carat gold is pure gold, 18-carat gold has 18 parts gold and 6 parts other materials. Pure gold is very soft. Adding another metal makes the jewellery harder and more durable.

Go to http://mypsci3.nelsonnet.com.au and click on **Metals and non-metals** to look at the different properties of metals and non-metals.

FIGURE 4.7 Bronze is a common alloy used to create sculptures, such as this famous work, called 'The Thinker', by Rodin.

Non-metals and metalloids

The elements found on the right-hand side of the zigzag line of the periodic table are non-metals. Elements such as helium (He), carbon (C), sulfur (S) and chlorine (Cl) are non-metals. The non-metal elements have a wide range of properties but, in general, they are poor conductors of heat and electricity, and are brittle as solids.

Hydrogen is an exception to the generalisations. It is found on the left-hand side of the periodic table, is a gas at room temperature and does not conduct electricity. This means that it is usually classified as a non-metal.

Neon is a non-metal located in group 18 of the periodic table. It has low melting and boiling points and therefore is a gas at room temperature. When a voltage is applied to the gas, a bright red light is emitted. This is used in neon lights. Helium, another group 18 non-metal, is also a gas at room temperature. It has a very low density, which means that it will rise above other gases. This is utilised in balloons that need to stay up in the air.

Some elements, the metalloids, also known as semimetals, fit somewhere between metals and non-metals because they have properties of both groups. These elements are found along the line dividing metals and non-metals on the periodic table. Boron, silicon, germanium, arsenic, antimony, tellurium and polonium are the metalloid elements.

Silicon is widely used in electronics as it is a semiconductor. This means that its electrical conductivity is between that of a conductor, such as a metal, and an insulator, such as a non-metal. Being a semiconductor means silicon parts of a circuit can very carefully control the electrical current through the circuit.

FIGURE 4.8 Helium balloons and a neon sign

REVIEW

1. Match the property with its meaning.
   - a Density
   - b Hardness
   - c Malleability
   - d Ductility
   - e Electrical conductivity
   - f Heat conductivity
   - i Ability to allow an electrical current to pass from one end to the other
   - ii Ability to be drawn into a wire
   - iii Ability to resist being scratched
   - iv Ability to pass heat from one end to another
   - v Ability to be beaten into another shape by a hammer
   - vi The mass of each cubic centimetre

2. Classify each of the following elements as metal, non-metal or metalloid.
   - a Lead
   - b Bromine
   - c Calcium
   - d Hydrogen
   - e Arsenic
   - f Neon

3. Are the following statements true or false? Rewrite the false statements to make them true. Be careful – some are tricky!
   - a Metals need to be heated over very hot flames to get them to melt.
   - b Dense metals are all very hard.
   - c There are more than 70 different metals on Earth.
   - d If you touch a live wire with a piece of any metal, you will get an electric shock.
   - e If something isn't shiny, it cannot be a metal.

4 Figure 4.9 shows three different ways metals are used. For each use, state three properties that the metal used must possess.

FIGURE 4.9 Three different uses for metals: (a) cooking pots over a campfire, (b) a sulfuric acid production plant, (c) a Russian scientific vessel bound for Antarctica

5 A gold atom is about eight times heavier than a magnesium atom. The density of gold is 19.3 g cm^{-3} and that of magnesium is 1.74 g cm^{-3}.
   a How many times more dense is gold than magnesium?
   b Suggest why gold is the denser metal.

How do metals react?

We can classify the properties of matter as physical properties or chemical properties. Physical properties are measurable without a change in composition. Melting point, density, colour and malleability are all physical properties. Chemical properties relate to a change in composition, or a chemical change. What a substance will or will not react with and how vigorous the reaction will be are chemical properties.

You have already learnt that group 1 and 2 metals are very reactive. In this section, you will compare the reactivity of different metals and look at specific reactions of metals.

ATL

ORGANISATION: SUPPORT SYSTEMS
In order to learn effectively, it is important to have a supportive environment. This includes having a quiet, organised place to work; developing a routine of completing homework, assessments and study; and having the support of your family and friends.

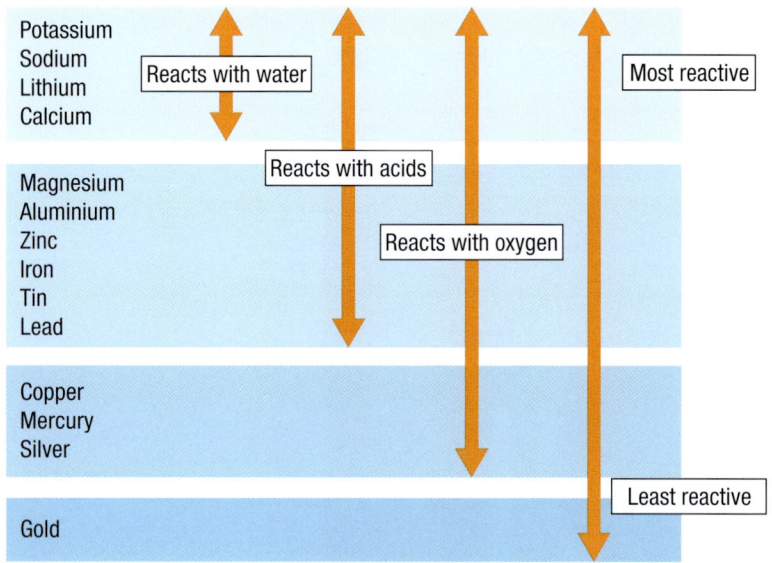

FIGURE 4.10 The reactivity series of metals

Reactivity series of metals

Metals vary significantly in how easily they will react. Metals such as gold and silver are very unreactive, which is why they are used for jewellery that we want to last for a long time. Others, such as sodium, are so reactive that they do not exist in their pure form in nature.

The **reactivity series of metals** is a list of metals arranged in their order of reactivity (Figure 4.10). It helps us to predict what a certain metal will react with. The least reactive metals, such as gold, will rarely react with anything. The metals that are a little more reactive, such as copper, will react only with some acids such as concentrated nitric acid. Then there are the metals such as magnesium and zinc that will also react with more dilute acids. The most reactive metals will even react with water.

Specific reactions of metals

There are certain general reactions that metals commonly undergo. You will learn about these in more detail in Unit 5, but we will briefly consider some of them now as they affect how different metals can be used.

Corrosion

Corrosion of metals occurs when the metal reacts with its environment, particularly oxygen in the air. This causes the solid metal to form **compounds** with the oxygen. Some unreactive metals, such as platinum, will not corrode, which is why they last a long time, and can be found as a solid metal in nature. Others, such as potassium, will react instantly.

FIGURE 4.11 Copper turns green when corroded.

The most commonly known example of corrosion is rusting, in which iron forms iron oxide or rust. Rusting occurs when iron is exposed to oxygen and water, especially salt water. This is why cars rust more quickly at the beach. Due to the huge impact of rust, various methods of preventing rust have been developed. Each of these methods works by preventing the chemical reaction. You may already be aware of some of these, such as painting and lubrication.

Not all metals corrode in the same way. We often think of aluminium as a metal that doesn't corrode. This is why it is used to make things such as window frames and cooking pots. In fact, pure aluminium reacts the instant it is placed in the air. However, the new substance it forms, aluminium oxide, is in the form of a transparent coating like a clear lacquer, so you don't realise that it is there. This layer is impermeable, meaning that substances are not able to pass through it. Therefore, the layer of aluminium oxide protects the metal underneath from further corrosion.

ACTIVITY: Rust prevention

Research the different methods of preventing the corrosion of iron. Identify how each method works and the situations where it would be the most effective. What are the advantages and disadvantages of each? How might factors such as economy, the environment, culture or ethics influence which method is implemented? Based on your research, choose the method that you believe is the best one to use and prepare a short presentation or report to present your decision and the evidence to support it. Make sure you keep a record of each source of information. These should be listed in alphabetical order in your bibliography.

INVESTIGATION 4.1: Investigating rust

YOUR CHALLENGE
To investigate the methods of preventing the rusting of iron nails.

THIS MIGHT HELP
Many nails found in hardware stores are made of iron and therefore are prone to rust. If these nails are placed in a solution containing gelatine and an indicator called potassium hexacyanoferrate(III), the indicator will turn dark blue in the areas where rust is occurring.

During your investigation you may treat plain iron nails with different rust prevention techniques. You may also use nails that have been pre-treated, such as galvanised nails.

In your planning, consider how you will control your variables to make this a fair test and how many trials you will conduct. You will also need to decide how you will measure or compare the amount of rust.

Once you have planned your method, get approval from your teacher and then conduct your investigation. Don't forget to keep a record of your data.

Carry out and write up your investigation following the guide in Appendix 3 on page 177 or as advised by your teacher. Refer to the descriptors for criteria B and C to make sure that you address all the areas needed.

Go to http://mypsci3.nelsonnet.com.au and click on **Preventing rusting** for a starting point for your investigation.

Tarnish

If you've ever cleaned silver jewellery or cutlery, then you have observed **tarnish**. Tarnish is a thin layer that forms at the surface of a metal when it reacts with substances found in the air. This is often oxygen, hydrogen sulfide or sulfur dioxide. Tarnishing is a self-limiting process, as once the layer is formed, it protects the metal below and no further reaction occurs. Some corrosion is considered tarnishing. For example, the corrosion of aluminium forms a self-protecting layer. Silver will tarnish if it is exposed to hydrogen sulfide in the air. This makes the silver atoms form the compound silver sulfide, which we see as a black tarnish.

FIGURE 4.12 A silver coin will turn black due to tarnish.

Go to http://mypsci3.nelsonnet.com.au and click on **Reactions with acids** to interact with a simulation of different metals reacting with an acid.

Metals reacting with acids

Many metals will react with acids to form a **salt** and hydrogen gas (H_2). The gas can be observed as bubbles escaping from the solution. As the metal reacts, it dissolves in the acid.

Unfortunately, in some locations, rain may be acidic due to the chemicals that dissolve in the water vapour. This means that, when it rains, the acidic water will fall on structures. If a metal structure reacts with the acid, it will dissolve. Therefore, in areas prone to acid rain, it is important to choose materials, including metals, that do not react with acids, or protect them from the rain.

Metals reacting with water

Some very reactive metals will also react with water. The metals in group 1 of the periodic table will react with water at room temperature. The metals in group 2 are not quite as reactive and vary in their reactions with water. Calcium, strontium and barium will react with water at room temperature. Magnesium is less reactive and will only react with water if the water is heated to make steam. Beryllium will not react with water at all.

During the reaction between metal and water, the metal will make a compound known as a hydroxide, for example, sodium hydroxide (NaOH). Hydrogen gas (H_2) is also formed and can be seen as bubbles in the solution.

When choosing a metal for a particular purpose, it is important to consider its physical and chemical properties. Imagine if calcium metal were chosen to make a boat. The metal would react with the water and dissolve!

Making electricity from metals

A huge advantage of the different reactivity of metals is that metals with different reactivities can be used to make electricity. All that is needed are two different metals (such as iron and copper) that are connected by wires, and a solution that conducts electricity, such as a salty or acidic solution. Even a potato or lemon can provide a suitable solution, as they are full of mineral salts.

The specific metals used to make a battery are not chosen randomly. The combination of metals will determine the voltage, or strength, of the battery. Factors such as availability, cost and safety must also be considered.

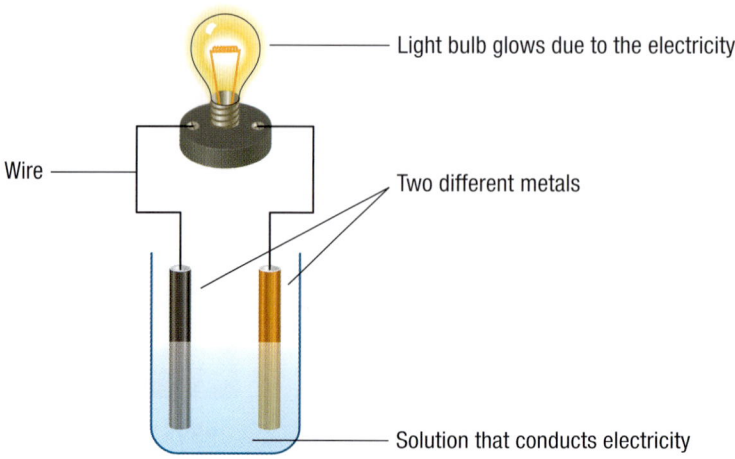

FIGURE 4.13 A chemical reaction producing electricity

EXPERIMENT 4.2 — A potato battery

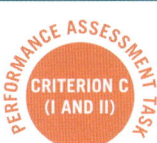

AIM
To make electricity using an iron nail, a strip of copper metal and a potato.

MATERIALS
- potato (unpeeled)
- two wires with connectors at each end
- large iron nail
- strip of freshly cleaned copper metal
- multimeter, light-emitting diode (LED), micro-ammeter or other current-detecting device

PROCEDURE
1. Insert the iron nail and copper strip into the potato about 1 cm apart. (They must not be touching!) Leave at least 1 cm of each hanging out to connect to the wire.
2. Attach a wire to each metal and connect the other ends of the wires to the multimeter or other device.

RESULTS
Record the current that the multimeter is showing. If no current is produced, check that all your contacts are good and try connecting the wires the other way around.

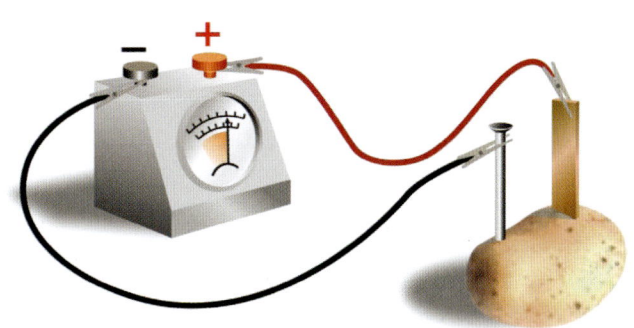

FIGURE 4.14 A potato battery

DISCUSSION
1. What did you observe when everything was connected?
2. Did your potato battery produce a current? Remember, any electrical current could only have come from the metals and the potato – you did not plug this into a power point!
3. What do you think would happen if you unclipped one of the wires? Check your prediction with your potato battery.
4. What do you think would happen if one of the metals was not in the potato? Use your potato battery to see if your prediction is correct. Why do you think this is so?
5. What do you think would happen if the two metals touched? Test your prediction. Can you provide an explanation for your observation?
6. If the potato is left for a long time, it dries out. What do you think would happen to the potato battery then? What knowledge did you base your answer on?

CONCLUSION
What did you conclude regarding the potato battery?

CRITICAL THINKING: HYPOTHESISING
Consider the characteristics of a good hypothesis, including the explanation for the hypothesis.

Making electricity

INVESTIGATION 4.2

YOUR CHALLENGE
To investigate the factors that affect the electrical voltage and current produced in a battery.

THIS MIGHT HELP
A battery needs two different metals and a chemical system, such as the body of the potato.

Think about what variables could affect the amount of electricity produced. Choose one of these variables to investigate. This will be your independent variable.

You will need to measure the current produced in your battery, the dependent variable. How will you measure this?

You will need to control all the other variables to make this a fair test.

- What will you use as your chemical system? You could use a potato, fruit, solutions of fruit juices, or even salt solutions or weak acids.
- What combinations of metals will you use?
- What size will your metals be?
- What concentrations of solutions will you use?
- How far apart will you place the metals?
- How will you connect the parts of the battery?

Carry out and write up your investigation following the guide in Appendix 3 on page 177 or as advised by your teacher.

DISPOSAL OF BATTERIES

We are becoming more and more reliant on technologies that require batteries. These batteries often contain substances that are potentially harmful if they are disposed of in the rubbish bin. These substances include heavy metals, which leak into the ground and build up in the soils, water and wildlife. It is important that batteries are recycled appropriately, and not placed in the normal recycling rubbish. Find out where you can dispose of batteries in your community. What action can you put into place in your home, school or community to ensure batteries aren't being dumped?

FIGURE 4.15 Some everyday objects that have batteries

Electroplating

When we make electricity from metals, we are using a chemical reaction to make the electricity. We can also do the reverse – use electricity to make a chemical reaction happen. During this chemical reaction, the metal in the compound can become pure metal. This means that a deposit of solid metal is formed. We are able to control the reaction, so it is possible to decide where the metal deposits. We use this type of reaction to put a layer of metal onto another surface in a process called **electroplating**. This produces items such as gold-plated jewellery and chromium-plated bull bars for cars and trucks.

FIGURE 4.16 Chrome plating protects the bull bars on trucks and other vehicles.

Go to http://mypsci3.nelsonnet.com.au and click on **Electroplating**. Use the animation to observe the nickel plating of a copper strip. Can you see how the part of the copper in the nickel solution changes colour as the nickel is plating?

EXPERIMENT 4.3 Electroplating

Electroplating uses electricity to deposit a layer of metal on an object. It is often used to protect the underside of the object, such as in the gold-plating of electronic devices or the chrome-plating of bull bars for cars and trucks. It is also used to make objects look more appealing, for example, gold-plated jewellery.

AIM
To electroplate a key with a layer of nickel (or copper) metal.

MATERIALS
- brass key
- 200 cm³ electroplating solution containing nickel (or copper)
- emery paper or steel wool
- glass rod
- strip of nickel (or copper)
- wooden peg
- 2-volt direct current (DC) source and electric leads
- switch
- 100 cm³ glass beaker
- 250 cm³ glass beaker
- 12 cm length of copper wire
- latex gloves
- plastic tray
- paper towel
- rinse bottle and distilled water
- stereomicroscope (optional)

>

SAFETY ADVICE

Although they are not as toxic as the solutions used to plate items with gold and silver, the solutions you use in this experiment require you to observe the following precautions.
- Wear gloves at all times during this experiment and wash your hands thoroughly afterwards with soap and water.
- Place the solutions, including the rinse water, in the residue jars provided ready for chemical treatment. Do not tip them down the sink.
- Tell your teacher immediately if any of the solution spills onto you or the bench.

PROCEDURE

1 Examine the key under the stereomicroscope and note its appearance.
2 Put on the gloves, then clean the brass key with the emery paper over the plastic tray.
3 Rinse the key with a small amount of distilled water over the 100 cm^3 beaker and then dry it with paper towel.
4 Set up your experiment as shown in Figure 4.17. Do not switch on the circuit until your teacher has checked it. Ensure you set the voltage at 2 volts.
5 Let the current flow for 20 minutes. Every now and then, gently swirl the key a little to mix up the solution. Record any changes to the solution, the key and the nickel or copper strip.
6 After the plating is complete, rinse the key again over the 100 cm^3 beaker and dry it with paper towel. Examine it under the stereomicroscope and record its appearance.
7 Place your electroplating solution and rinse water in the residue jar provided and pack up according to your teacher's instructions. Remove your gloves and wash your hands thoroughly.

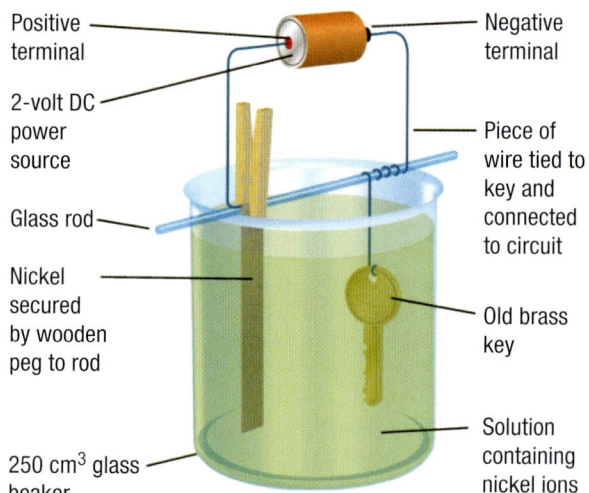

FIGURE 4.17 The correct set-up for nickel plating. If you are aiming to copper plate the key, use a strip of copper and a solution containing copper ions.

RESULTS

Record your results, including your observations, in a table.

DISCUSSION

1 What was the evidence that the brass key was successfully nickel or copper plated?
2 Was there any evidence to suggest that the brass or its coating was composed of crystals? Discuss this, referring to your results.
3 Discuss whether the conditions and solution you used would be suitable for commercial plating.
4 Discuss whether there was any evidence to suggest that the strip of metal was involved in the reaction.
5 Examine the material safety data sheet for the nickel (or copper) compound. What does it tell you about what to do if you get any of the solution on you?

CONCLUSION

What conclusion can you make regarding the electroplating of a key?

EXTENSION

1 Predict what might have happened had you continued to let the current flow. Would the plating have simply kept growing?
2 A direct current (DC) flows in only one direction along the wires. An alternating current (AC) alternates between flowing for a fraction of a second in one direction and then for a fraction of a second in the reverse direction. Suggest why you used a DC and not an AC power source for the electroplating.
3 What do you think might happen if you used a higher voltage?

REVIEW

1 Place the following metals in their order of reactivity: potassium, zinc, magnesium, copper, gold, iron.
2 State whether you would expect copper to react with water. Justify your answer.
3 Describe the difference between corrosion and tarnish.
4 Explain how painting the surface of an iron structure helps to reduce rust.
5 Explain why magnesium ribbon (a thin strip of magnesium) often has a black colour on its surface.
6 Vinegar (acetic acid) is stored in plastic or glass bottles, not metal containers. Explain what might happen if it was stored in a metal container.
7 Use the terms 'chemical reaction' and 'electricity' to explain the difference between electroplating and a battery.
8 Label the parts needed to make electricity from metals in the diagram below.

FIGURE 4.18 Metals producing electricity

UNIT QUESTIONS

CRITERION A

EXPLAINING SCIENTIFIC KNOWLEDGE

1. Classify each of the following substances as an alkali metal, an alkaline earth metal, a transition metal, another metal, a non-metal, a metalloid or an alloy. (Level 1–2)
   a. Barium
   b. Fluorine
   c. Brass
   d. Argon
   e. Copper
   f. Aluminium
   g. Lithium
   h. Silicon

2. State the products from the reaction of:
   a. a reactive metal and an acid
   b. an alkali metal and water
   c. a metal and the oxygen in the environment. (Level 3–4)

3. Describe what each of the following terms mean. (Level 5–8)
   a. Malleable
   b. Opaque
   c. Corrosion
   d. Reactivity series of metals

4. Describe how you could produce electricity from a strip of copper and an iron nail. State what other equipment you would need and draw a simple diagram to show your experimental set-up. (Level 5–8)

APPLYING SCIENTIFIC KNOWLEDGE AND UNDERSTANDING TO SOLVE A PROBLEM

5. For each of the following uses of metals, identify the important properties that are being applied. (Level 1–4)
   a. Mirrors in medical lasers
   b. High-voltage power lines
   c. Engine nozzle of a jet aircraft
   d. Fuel storage tanks
   e. Building towers
   f. Pinning a broken bone

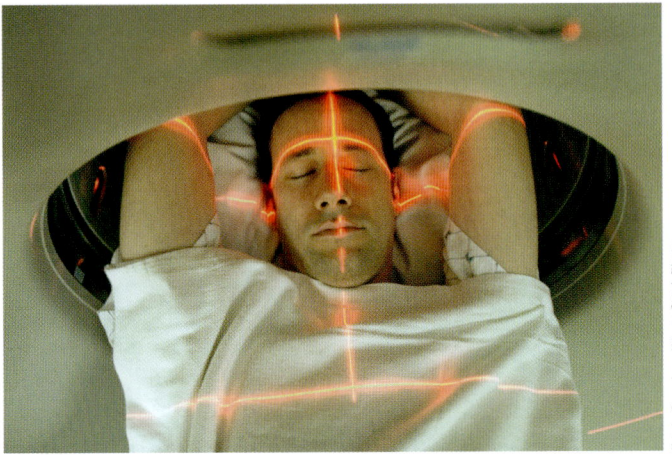

FIGURE 4.19 Medical lasers need to produce high-energy beams that can be controlled very precisely.

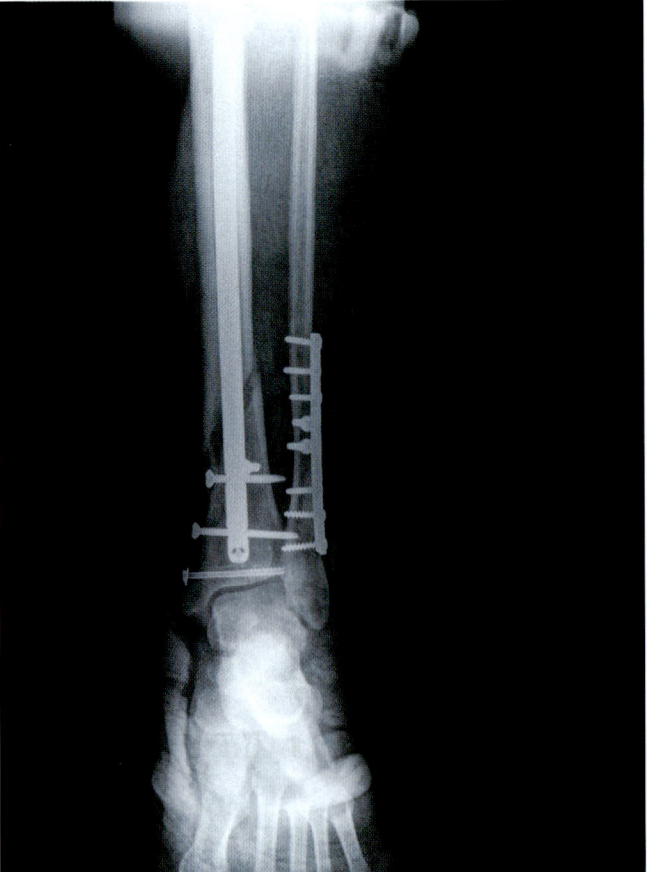

FIGURE 4.20 A broken leg, mending with the assistance of metal plates and pins

6   Part of the reactivity series of metals is shown below.
    caesium, potassium, calcium, aluminium, zinc, iron, nickel, tin, lead, copper, silver
    a   Predict what caesium might do if you added a small piece of it to water. Justify your answer. (Level 1–4)
    b   State which of the metals in the list is the least reactive. (Level 1–4)
    c   Discuss which of the metals in the list appears to be far less reactive than it really is. (Level 1–4)
    d   Explain why a coating of tin helps protect steel from corrosion. (Level 1–4)
7   Explain the advantages and disadvantages of gold-plated jewellery compared to pure gold jewellery. (Level 1–4)
8   A scientist discovers a new element that is a solid at room temperature, will conduct electricity, does not react with water and forms a pink compound. Outline where you would recommend the scientist place the element on the periodic table. Justify your recommendation. (Level 5–8)
9   Metals make up a large proportion of all the elements. Describe how life would be different if all the elements were metals. (Level 5–8)
10 Which metals do you think were the first to be discovered? Which were the last to be discovered? Justify your answers using your knowledge of metals. (Level 5–8)
11 Suppose that humans had not discovered how to obtain metals from rocks, and the only metal available for use was gold. How would your life be different? (Level 5–8)
12 Metals vary in their properties and therefore what they are used for. Use your knowledge of metals and their properties and uses to choose which metal is the best. Provide evidence to support your choice. (Level 1–8)

INTERPRETING INFORMATION

13 The pie graphs in Figure 4.21 show the composition of the Earth's crust.
    a   State which element is the most abundant in the Earth's crust. (Level 1–4)
    b   Calculate the percentage of the Earth's crust that is made of metals, assuming that the 'Others' are metals. (Level 1–4)

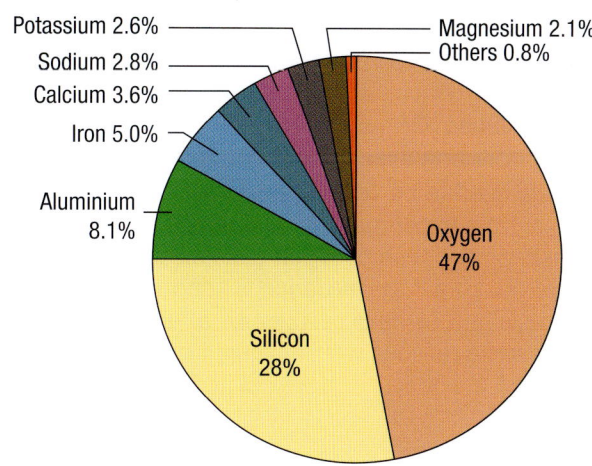

FIGURE 4.21 The composition of the Earth's crust

   c   Use the data in Table 4.4 to draw a pie graph for the elements found in the Earth's atmosphere. (Level 1–2)

TABLE 4.4 The composition of the Earth's atmosphere

Element	Percentage abundance
Nitrogen	78
Oxygen	21
Argon	0.93
Carbon	0.03
Neon	0.0018
Helium	0.00052

   d   Discuss why there are no metallic elements in the six most abundant elements in the atmosphere. (Level 5–8)

14 In etching, a waxy layer is placed on the surface of a metal plate. The artist cuts through the waxy layer, making the pattern or picture. This is then painted with, or dipped in, acid that 'eats out' the metal wherever the wax has been cut. The wax is then removed and the picture is displayed where the acid has reacted with the metal. Use your knowledge about the reactions of metals to explain how this process works. Would it work with all metals? (Level 5–8)

15 Metals are used in a variety of ways in medicine. Reflect on what you have learnt in this unit, and write a persuasive paragraph responding to the statement that 'metals have improved the quality of life'. (Level 5–8)

REFLECTION

1. The key concept in this unit was relationships. Explain how the term 'relationships' is used in scientific experimentation. Give some examples from work carried out during this unit.
2. Scientists need to be very careful in how they use evidence from experiments to claim they have established relationships. What does a scientist do if they are not sure if the evidence is adequate to allow them to claim they have established a relationship? Suggest some reasons why this might happen in an experiment.
3. Explain how the concepts of form and function apply to the use of metals.
4. Suggest some consequences of iron rusting in machines and cars. Why can't the problem of rusting be solved?
5. The Incas referred to gold as the 'tears of the Sun'. Why do you think gold has always fascinated humans since ancient times?

UNIT 5

USEFUL CHEMICAL REACTIONS

KEY CONCEPT
Change

RELATED CONCEPTS
Systems

Models

Conditions

GLOBAL CONTEXT
Scientific and technical innovation – an exploration into different chemical reactions that are important in our lives

STATEMENT OF INQUIRY
Modelling chemical changes that occur within a chemical system allows scientists to predict the consequences of combining specific chemicals.

INQUIRY QUESTIONS

FACTUAL
1 What is the difference between a physical change and a chemical change?
2 What is the law of conservation of mass?

CONCEPTUAL
3 How do balanced chemical equations model chemical reactions?
4 How can we classify chemical reactions?

DEBATABLE
5 Can we predict all chemical reactions?

Introduction

Our lives are filled with chemical reactions. Our bodies use chemical reactions to break down food, and then other chemical reactions use this food and oxygen to give us energy. Cooking food involves chemical reactions, as does running a car. Even the effect of sunlight on our skin is a chemical reaction.

There are many advantages to chemical reactions. But there are also some disadvantages. Some reactions produce undesirable substances that can cause significant problems. As citizens, it is important that we consider the impact of chemical reactions in which we are involved and take responsible actions.

Useful chemicals in our lives

Design a double-page layout for a popular science book for students of your age about chemicals in our lives.

Your double-page layout should be based on one of the categories of chemicals studied in this unit.
- Acids
- Bases
- Metals
- Carbonates
- Hydrogen carbonates (or bicarbonates)
- Salts

Do some further research into the uses of one of these types of chemical. Also do some research into the layouts used in books of this kind. One possible series of books to look at is the DK Eyewitness series (you will find sample pages on their website).

COMMUNICATION

Effective and correct use of scientific language. A crucial skill when writing a textbook for students of your age is to be able express ideas clearly and in ways appropriate to the age and knowledge of the students.

What do you already know about chemical reactions?

ACTIVITY

You have already learnt a lot about chemical reactions. Collect a number of sticky notes or small flash cards. Write one thing that you know about chemical reactions onto each one. Arrange them in a manner that shows how the different ideas are linked together. This makes a concept map. If you are not familiar with concept maps, your teacher will provide you with some instructions. When you are happy with your arrangement, glue the notes or cards onto a large piece of paper. As you progress through the unit, look back at your concept map and add new ideas.

CRITICAL THINKING

Ability to use concept maps or similar to organise your ideas about a topic

Chemical and physical change

Changes in chemicals can be classified as either physical or chemical. The difference is based on whether products are formed. If new products are formed, then it is a **chemical change**, or **reaction**. If a new substance has not been formed, then it is a **physical change**.

A physical change often involves a change of **state**. Boiling water, dissolving sugar and crushing a solid are all examples of physical changes.

If a chemical reaction occurs, the products formed have different properties from the chemicals you started with. This means that you may observe a colour change, a solid or **precipitate** being formed, a change in temperature or a gas being produced and seen as bubbles. Wood burning, cooking a cake, bread rising and iron rusting all involve chemical reactions.

FIGURE 5.1 Dissolving sugar is a physical change.

FIGURE 5.2 Burning wood is a chemical reaction.

EXPERIMENT 5.1 Observing chemical reactions

AIM
To observe evidence of a chemical reaction.

MATERIALS
- 250 cm^3 beaker
- Two 100 cm^3 beakers
- sodium thiosulfate solution, 1 mol/dm^3, approximately 1 cm^3
- acidified potassium permanganate solution, 0.0025 mol/dm^3, approximately 50 cm^3
- magnesium sulfate solution, 0.5 mol/dm^3, approximately 25 cm^3
- sodium carbonate solution, 0.5 mol/dm^3, approximately 25 cm^3
- citric acid
- sodium hydrogen carbonate (bicarb soda or baking soda) solid

SAFETY
For all experiments, wear eye protection and take care when handling chemicals. Wash up immediately any chemicals that are spilt on the desk or your hands.

PROCEDURE
Part A: Colour change
1. Place approximately 1 cm^3 of sodium thiosulfate into a 100 cm^3 beaker.
2. Pour approximately 50 cm^3 of acidified potassium permanganate solution into the same beaker.
3. Record your observations in your results table.

>

Part B: Producing a solid or precipitate
1. Place approximately 25 cm³ of sodium carbonate solution into a clean 100 cm³ beaker.
2. Pour approximately 25 cm³ of magnesium sulfate solution into the same beaker.
3. Record your observations in your results table.

Part C: Producing a gas and a change in temperature
1. Place 1 teaspoon of citric acid and 1 teaspoon of sodium hydrogen carbonate into the 250 cm³ beaker.
2. Add enough water to cover the solids and mix.
3. Use a thermometer to observe what happens to the temperature of the mixture.
4. Record your observations in your results table.

RESULTS
Use a table to record your observations.

DISCUSSION
For each reaction, describe the evidence that a chemical change occurred.

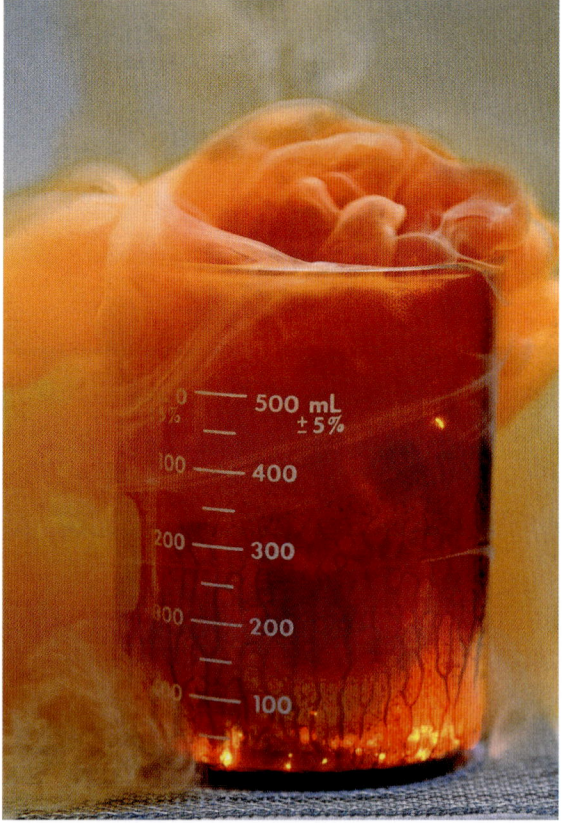

FIGURE 5.3 Evidence of chemical reactions

Gases produced in chemical reactions

Hydrogen (H_2), oxygen (O_2) and carbon dioxide (CO_2) are all colourless, odourless gases that are produced by many common reactions. They all look and smell the same, so to tell them apart we have specific tests that allow us to identify them.

Hydrogen gas

If a lit **splint** or **taper** is placed near hydrogen gas, a 'pop' sound can be heard. This is called the **pop test**.

Oxygen gas

A glowing splint or taper is used to identify oxygen gas. If the glowing splint or taper is held near the oxygen gas, it will re-light, producing a flame again.

Carbon dioxide

The **limewater test** is used to identify carbon dioxide. **Limewater** is actually a solution of calcium hydroxide ($Ca(OH)_2$). When carbon dioxide is bubbled through the limewater, the solution turns milky white. This is due to the carbon dioxide reacting with the limewater producing a precipitate (solid) of calcium carbonate ($CaCO_3$).

FIGURE 5.4 A glowing splint or taper will re-light in the presence of oxygen gas.

FIGURE 5.5 The limewater test

REVIEW

1. Classify each of the following as either a physical or a chemical change.
   a. Dissolving salt in water
   b. An aspirin tablet fizzing when it is added to water
   c. Burning a piece of meat
   d. Ice cream melting when it is left out of the freezer
2. Match the gas with the test used to identify it.

Gas	Test
Oxygen gas	Pop test
Hydrogen gas	Limewater test
Carbon dioxide	Glowing splint will re-light

3. Describe how a physical change differs from a chemical change if you were able to see the individual atoms or molecules.
4. In each of the photos in Figure 5.6, what evidence of a chemical reaction are you able to observe?

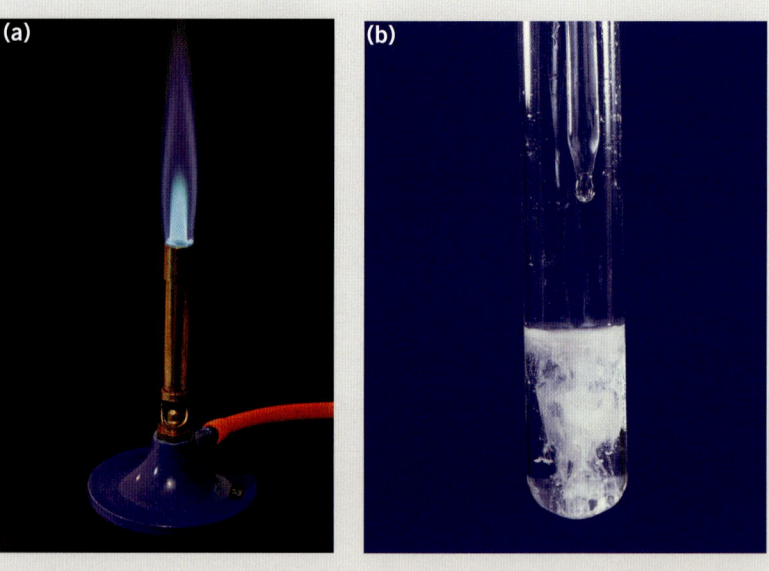

FIGURE 5.6 Chemical reactions

Representing chemical reactions

We are not able to actually see what is happening during a chemical reaction at an atomic level. Therefore, scientists have developed methods of modelling the chemical system to represent the chemicals involved and how they interact and change during the reaction.

One way of modelling the reaction is by using molecular models. These typically use balls of different colours to represent **atoms**. Rearranging the balls, or atoms, represents the rearrangement of atoms during a chemical reaction to form new chemicals.

Equations are another way to model reactions. In an equation, words or chemical symbols are used to represent the chemicals.

Modelling chemical changes

The chemicals that are present at the start of a chemical reaction are called **reactants**. These react to form the **products** of the reaction.

We often represent a chemical reaction with an equation. The reactants are written first. An arrow represents the progress of the reaction, then the products are written.

The reactants and products may be written as words to form a **word equation**, or as their chemical symbols to form a chemical or symbol equation.

When magnesium burns, it reacts with oxygen gas in the air, producing magnesium oxide. In this chemical reaction, magnesium and oxygen are the reactants and magnesium oxide is the product. This reaction can be represented by the equations shown below.

Word equation:

    magnesium + oxygen gas → magnesium oxide
            reactants   product

Chemical or symbol equation:

    $2Mg(s) + O_2(g) \rightarrow 2MgO(s)$
        reactants   product

The numbers in front of the chemical symbols are called **coefficients**. These show how many **molecules**, or atoms, of each chemical are needed or produced. It is a little like saying you need two pieces of bread and one slice of cheese to make a sandwich.

There are also letters written in brackets after the symbols. These show the **state** of the substance – solid (s), liquid (l) or gas (g). Sometimes you may see (aq) as the state. This stands for 'aqueous' and means that it is dissolved in water.

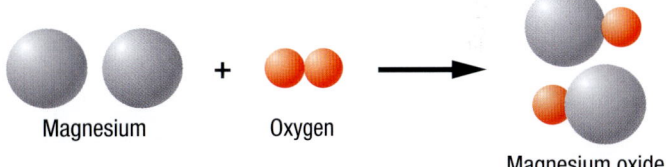

FIGURE 5.7 Models showing the reaction between magnesium and oxygen

TABLE 5.1 The letters representing the states of matter

Letter	State
s	solid
l	liquid
g	gas
aq	aqueous

Balancing chemical equations

In the late 18th century, Antoine Lavoisier discovered that the mass of a chemical system remains the same during a chemical reaction. This became known as the **law of conservation of mass**, which states that mass is neither gained nor lost during a chemical reaction. This means that the atoms that were present at the start of the reaction are also there at the end. However, during the reaction they may have been rearranged.

Using the law of conservation of mass, we are able to write a **balanced chemical** or **symbol equation**. This shows the number of each element or compound involved in a chemical reaction. When an equation is balanced, the number of atoms of each element in the reactants is the same as the number of atoms of each element in the products.

Follow these steps to write a balanced chemical or symbol equation.

1. Write the reactants and products using their chemical symbols, separated by an arrow. This is known as a skeletal equation as it shows only what is involved in the reaction.
2. Choose one element. Count how many atoms of that element are on the reactant side, and how many are on the product side of the equation.
3. If the number of atoms is not the same, write a coefficient (number) in front of the appropriate chemical symbol so that they will become equal. Do not change the formula of the compounds when you balance the equation.
4. Work systematically through each element involved in the reaction, repeating steps 2 and 3 until the atoms of all the elements are balanced.
5. Do a final check to make sure that all elements are still balanced. Sometimes when you balance one element, it changes another one as well.
6. Write the state of each substance in brackets after the chemical symbol.

For example:

When methane is burnt in air it reacts with oxygen, producing carbon dioxide and water. The skeletal equation is:

$$CH_4 + O_2 \rightarrow CO_2 + H_2O$$

Working through each element separately:

FIGURE 5.8 Burning methane

- There is one carbon atom in the reactants and one in the product. Therefore, it is already balanced.
- There are four hydrogen atoms in the reactants and two in the products. Therefore, write a two in front of the H_2O. This now makes four hydrogen atoms in the products as well.
- There are two oxygen atoms in the reactants and four in the products. Therefore, write a two in front of the O_2. This now makes four oxygen atoms in the reactants as well. The final balanced equation is:

$$CH_4 + 2O_2 \rightarrow CO_2 + 2H_2O$$

- To complete the equation, write the state of each substance in brackets after its formula:

$$CH_4(g) + 2O_2(g) \rightarrow CO_2(g) + 2H_2O(g)$$

ACTIVITY: Balancing chemical equations

Use the weblinks to go to interactive sites for balancing equations.

Go to http://mypsci3.nelsonnet.com.au and click on **Balancing chemical equations.** Click 'start' and balance the equations.

REVIEW

1. State the law of conservation of mass.
2. Describe how we represent the state of matter for each substance involved in a chemical reaction.
3. In the following equation, clearly identify the reactants, the products, the states of matter and the coefficients used to balance the equation.
   $K_2O(s) + H_2O(l) \rightarrow 2KOH(aq)$
4. Balance each of the following skeletal equations.
   a. $H_2 + O_2 \rightarrow H_2O$
   b. $KI + Cl_2 \rightarrow I_2 + KCl$
   c. $CaCl_2 + AgNO_3 \rightarrow AgCl + Ca(NO_3)_2$
   d. $Al + O_2 \rightarrow Al_2O_3$
5. When a piece of chalk ($CaCO_3$) is added to a solution of hydrochloric acid (HCl), it dissolves, producing a solution of calcium chloride ($CaCl_2$) and water (H_2O). Bubbles of carbon dioxide (CO_2) are also produced. Represent this reaction with a word equation and a balanced chemical or symbol equation that includes the state of each substance.

Go to http://mypsci3.nelsonnet.com.au and click on **More balancing equations.** Click on 'Game' and then 'Level 1'. Balance the equations and then move onto the harder levels.

REFLECTION: MULTIPLE INTELLIGENCES

You will have noticed that not everyone thinks or learns in the same way. Howard Gardner's multiple intelligences theory describes seven different intelligences: linguistic, logical-mathematical, musical, bodily-kinaesthetic, spatial-visual, interpersonal and intrapersonal. Understanding your own preferred intelligence can help you utilise these strengths in your learning. Use the weblink to find out your natural intelligence.

Types of chemical reactions

Chemical reactions can be classified in a few different ways. In this book we are going to look at one method. You will learn about other methods in *Chemistry 4/5 for the international student*.

The method that you will consider here is looking at what happens to the reactants to become products. This will enable you to see patterns in the reactions, which will help you to predict the products for some reactions.

Synthesis reactions

During a **synthesis reaction**, reactants combine to form a single product. This means that there is more than one reactant and only one product. For example, in a fuel cell, hydrogen reacts with oxygen. The hydrogen gas (H_2) and oxygen gas (O_2) combine to form water (H_2O). This is represented in the following equation.

Word equation:

   hydrogen gas + oxygen gas → water

Chemical or symbol equation:

   $2H_2(g) + O_2(g) \rightarrow 2H_2O(l)$

Go to http://mypsci3.nelsonnet.com.au and click on **Howard Gardner's Multiple Intelligence tests.** Scroll down until you get to the heading 'multiple intelligence tests' and then choose one to try.

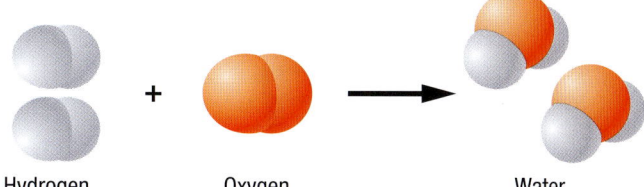

FIGURE 5.9 Model representing the synthesis reaction between hydrogen and oxygen

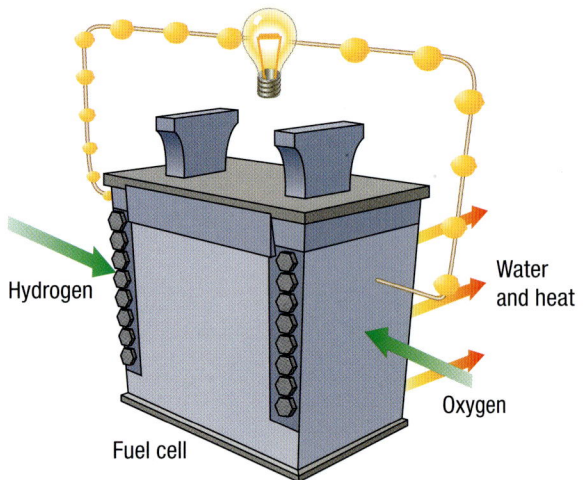

FIGURE 5.10 Hydrogen fuel cells use hydrogen and oxygen to produce electricity, heat and water, and are used to power some electric motor vehicles.

A synthesis reaction that occurs naturally in the atmosphere is the reaction between carbon dioxide (CO_2) and water (H_2O). These combine to produce carbonic acid (H_2CO_3). This is why even unpolluted rain is slightly acidic.

Word equation:

carbon dioxide + water → carbonic acid

Chemical or symbol equation:

$$CO_2(g) + H_2O(l) \rightarrow H_2CO_3(aq)$$

Understanding this synthesis reaction has helped scientists understand some of the effects of the increased levels of carbon dioxide in our atmosphere. One of these consequences is the oceans becoming more acidic due to production of carbonic acid. The increased acidity has affected the marine life as many organisms rely on certain environmental conditions, including acidity.

Decomposition reactions

Decomposition reactions are the opposite of synthesis reactions. During decomposition reactions a single reactant breaks up into smaller parts. This means that there is one reactant but more than one product. For example, when copper carbonate ($CuCO_3$) is heated it will decompose into copper oxide (CuO) and carbon dioxide (CO_2).

Word equation:

copper carbonate → copper oxide + carbon dioxide

Chemical or symbol equation:

$$CuCO_3(s) \rightarrow CuO(s) + CO_2(g)$$

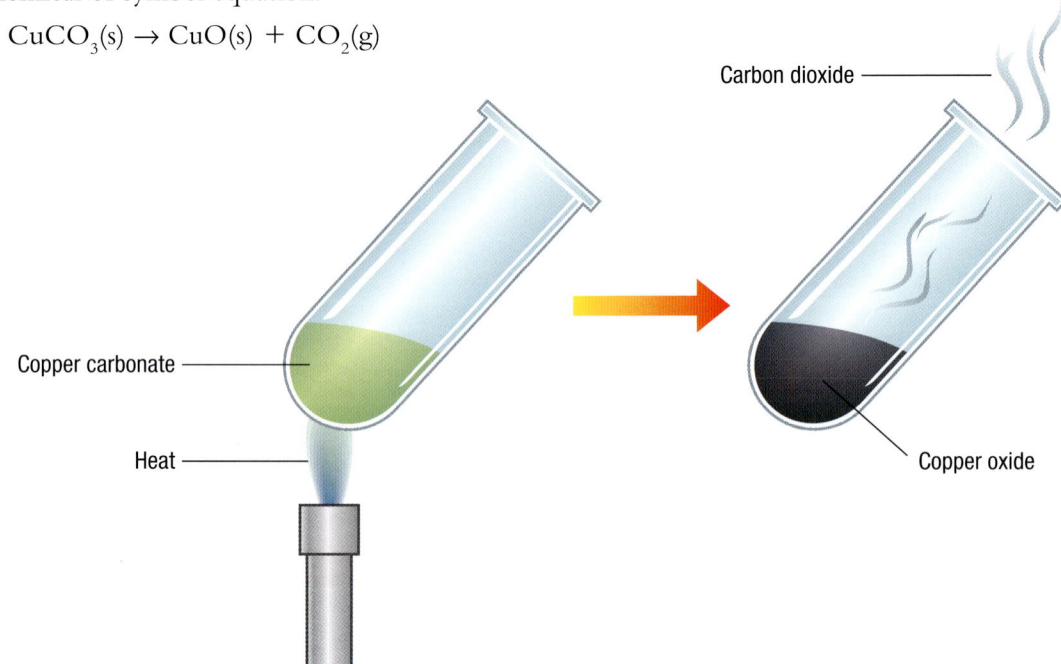

FIGURE 5.11 The decomposition of copper carbonate

Decomposition reactions are commonly used in cooking. Baking powder, which contains sodium hydrogen carbonate (NaHCO$_3$), is often added when making cakes. In the oven, the heat makes the sodium hydrogen carbonate decompose. The products of this reaction are sodium carbonate (Na$_2$CO$_3$), carbon dioxide (CO$_2$) and water. The carbon dioxide gas helps make the cake rise, making it light and fluffy:

$$2NaHCO_3(s) \rightarrow Na_2CO_3(s) + CO_2(g) + H_2O(g)$$

EXPERIMENT 5.2 Heating copper carbonate

AIM
To observe the decomposition of copper(II) carbonate

MATERIALS
- solid copper(II) carbonate (CuCO$_3$)
- 10 cm³ of limewater (Ca(OH)$_2$)
- two large test tubes
- stopper and delivery tube
- spatula
- Bunsen burner
- matches
- clamp and stand

SAFETY ADVICE
- Always wear safety glasses and do not inhale any of the dust from the solid.
- Do not dispose of any chemicals down the sink.

PROCEDURE
1. Place about 10 cm³ of limewater into a test tube.
2. Place a large spatula of copper carbonate into the second test tube and put the stopper with the delivery tube into the end of this test tube.
3. Secure the test tube containing the copper carbonate in the clamp and place the end of the delivery tube into the limewater, as shown in Figure 5.12.
4. Light the Bunsen burner, remembering to have the air hole closed when you light it.
5. Turn the Bunsen burner to a blue flame. Heat the copper carbonate gently at first, and then more strongly.
6. Record any observations you make.
7. When the reaction has finished, turn off the Bunsen burner and remove the delivery tube from the limewater.
8. Follow your teacher's instructions regarding the disposal of your chemicals.

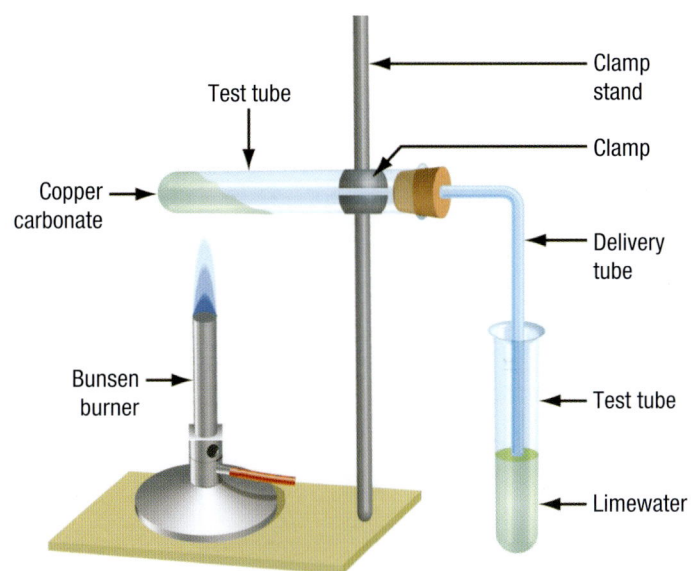

FIGURE 5.12 Experimental set-up

DISCUSSION
1. Write a word equation for the reaction that occurs.
2. Find out the formula for copper(II) oxide, and use this to help write a balanced chemical equation for the reaction.

3 Describe why this reaction is classified as a decomposition reaction.
4 Describe the observations you made that confirmed a chemical reaction had occurred.
5 Describe how you identified the gas as carbon dioxide.

CONCLUSION

Describe the conclusion that you are able to make regarding the decomposition of copper(II) carbonate.

EXTENSION

1 Do you think all carbonates would undergo a similar decomposition reaction upon heating?
2 How could you test your answer? Your teacher may provide you with chemicals to see whether your prediction is correct.

Displacement reaction

During a **displacement reaction** one part of a compound is displaced (or replaced) by another element. In essence, the two elements swap places. These can be single displacement reactions, in which a single compound is affected, or double displacement reactions, which involve two compounds. When we write the equations for these reactions, we simply swap the relevant elements to write the products.

FIGURE 5.13 A silver chloride solid forms in the solution when silver nitrate and sodium chloride are mixed.

For example, if solutions of silver nitrate ($AgNO_3$) and sodium chloride (NaCl) are mixed, a white solid of silver chloride (AgCl) appears in a solution of sodium nitrate ($NaNO_3$). The silver and sodium have swapped. As there are two compounds involved, this is a double displacement reaction.

Word equation:

silver nitrate + sodium chloride → silver chloride + sodium nitrate

Chemical equation:

$AgNO_3(aq) + NaCl(aq) → AgCl(s) + NaNO_3(aq)$

Displacement reactions are used in preparing water for household use. In many situations, ground water is treated to make it suitable for drinking and use in our homes. This water often contains magnesium ions, which will form scum in kettles, hot water systems and other appliances. It also reduces the ability of soaps to lather. In order to remove the magnesium ions, lime (calcium hydroxide ($Ca(OH)_2$)) is added to the water. The magnesium displaces the calcium to form solid magnesium hydroxide ($Mg(OH)_2$). The solid can be removed by simple filtration methods.

Word equation:

magnesium ion + calcium hydroxide → calcium ion + magnesium hydroxide

Chemical equation:

$Mg^{2+}(aq) + Ca(OH)_2(aq) → Ca^{2+}(aq) + Mg(OH)_2(s)$

EXPERIMENT 5.3 — A metal displacement reaction

AIM
To observe a metal displacement reaction.

MATERIALS
- stereomicroscope
- three Petri dishes
- two small iron nails
- short piece of copper wire
- tight-fitting latex gloves
- dropper bottle of 0.1 mol/dm³ silver nitrate ($AgNO_3$)
- dropper bottle of 2 mol/dm³ copper(II) chloride ($CuCl_2$)

EXTENSION
- iron nail
- copper wire
- strip of zinc metal
- dropper bottle of 0.1 mol/dm³ iron(II) nitrate ($Fe(NO_3)_2$)
- dropper bottle of 0.1 mol/dm³ zinc(II) nitrate ($Zn(NO_3)_2$)

SAFETY ADVICE
- Silver nitrate causes permanent stains and will sting broken skin. Copper(II) chloride is toxic. While handling these chemicals, use tight-fitting latex gloves.
- Do not tip the mixtures down the sink: place them in the residue jars provided.

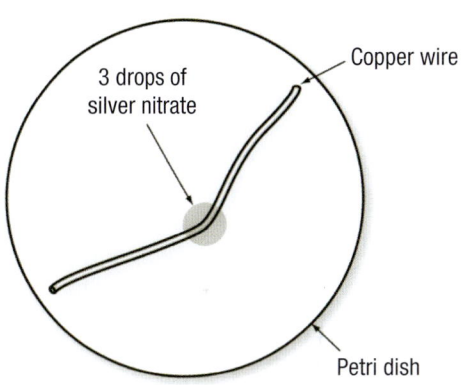

FIGURE 5.14 Set-up of the wire, Petri dish and silver nitrate

PROCEDURE
1. Place the copper wire in the centre of a Petri dish and focus on it under the microscope. Note the appearance of the wire.
2. Add three drops of silver nitrate to the centre of the wire, as shown in Figure 5.14, and observe all the changes that occur over the next 5 minutes. Keep this dish to the side and observe any further changes.
3. Repeat steps 1 and 2 with an iron nail and the copper(II) chloride solution.
4. Predict what might happen if you now add three drops of silver nitrate to an iron nail instead.
5. Try this combination and observe whether your prediction was correct.

RESULTS
Prepare a table to clearly show the metal, the solution and the observations.

DISCUSSION
1. What do you think the crystals might be?
2. In the first case, the silvery crystals are crystals of pure silver. The blue colour that eventually appears in the solution is due to the production of copper nitrate.

>

The word equation for the reaction is:

copper + silver nitrate → silver + copper nitrate

This shows us that the copper and silver have 'swapped places', or copper has displaced silver. Write the word equations for any other reactions you observed.

CONCLUSION

What conclusion can you make about the displacement of metals? Were your predictions correct?

EXTENSION

Predict what might happen if you placed a piece of copper wire in a solution of an iron compound, instead of an iron nail in a solution of a copper compound. State your reasoning. Your teacher will give you the chemicals needed for you to test whether your prediction is correct. Is it?

Predicting reactions

INVESTIGATION 5.1

YOUR CHALLENGE

Look at the reactivity series of metals in Figure 4.10 on page 76. What do you notice about the activity of the metal that swaps places with the metal in the compound? Use this to predict what you might see if you were to add an iron nail to a solution of zinc compound and a piece of zinc to a solution of an iron compound. State your reasoning.

THIS MIGHT HELP

Your teacher will provide you with the chemicals needed for you to test whether your predictions are correct. Are they?

Carry out and write up your investigation following the guide in Appendix 3 on page 177 or as advised by your teacher.

Combustion reactions

When a chemical reacts with oxygen and produces heat and light, it is a **combustion reaction**. The **combustion** of a metal, such as magnesium, is also a synthesis reaction. Another very common combustion reaction is the burning of carbon compounds such as petrol or **natural gas** (methane). These reactions produce carbon dioxide and water.

For example, the burning of methane (CH_4) in a gas cooker can be represented by the following equations.

Word equation:

methane + oxygen gas → carbon dioxide + water

Chemical equation:

$$CH_4(g) + 2O_2(g) \rightarrow CO_2(g) + 2H_2O(g)$$

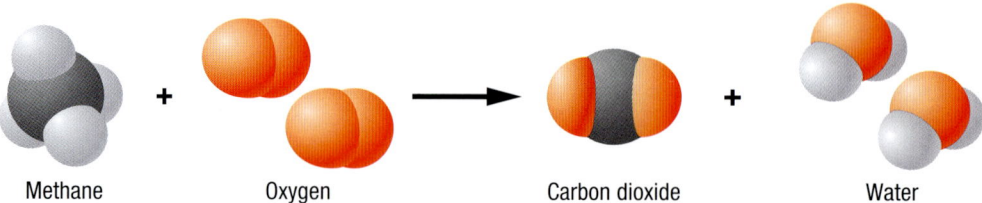

Methane   Oxygen   Carbon dioxide   Water

FIGURE 5.15 The combustion of methane

TA SAFETY WITH CARBON MONOXIDE

When there is not enough oxygen to produce carbon dioxide in the combustion reaction of carbon compounds, carbon monoxide is produced. This often happens in cars, heaters and stoves running in enclosed spaces. Carbon monoxide is highly poisonous. It prevents oxygen transport in the blood, and can result in death. Carbon monoxide is a colourless, odourless, tasteless gas and so may not be detected until it is too late. Cars are now fitted with catalytic converters, which convert much of the harmful emissions into less harmful substances. Talk about this risk with your family and friends. Discuss how you can ensure the safety of yourself and others.

Combustion reactions contribute many positive and negative aspects in our lives. Cars run by the combustion of petrol (or gasoline), gas or diesel. The combustion of coal is used to generate electricity. Many homes rely on the combustion of natural gas for heating and cooking. While these all enhance our lives, these combustion reactions also produce products that are released into the environment, causing pollution.

REVIEW

1. Match the type of reaction with its definition.
   - a Combustion
   - b Decomposition
   - c Synthesis
   - d Displacement
   - i One chemical displaces or swaps with another one in a compound.
   - ii Two or more reactants combine to produce a single product.
   - iii One reactant reacts with oxygen, producing heat and light.
   - iv A single reactant breaks down into more than one product.
2. What type of reaction takes place in fireworks?
3. During the process of digestion, large protein molecules are broken down into smaller amino acids. How would you classify this reaction?
4. Aluminium metal will burn in air, producing a brilliant white light. In this reaction, aluminium (Al) reacts with oxygen (O_2) to form aluminium oxide (Al_2O_3). Explain how this reaction can be classified as two different types of reactions.

More specific reactions

We can further classify chemical reactions by the types of reactants involved in them. This makes it easier to predict the products that are formed. Each of the reactions we will discuss here fits into one of the types mentioned earlier. Can you work out which? We will look at some reactions that you are more likely to observe, or conduct, in your school laboratory. Before we look at the reactions, we will revise what some types of reactants are.

Types of reactants

In *Science 1 for the international student* Unit 4, you learnt about **acids**, **bases** and metals. In this section, you will also look at their reactions with **carbonates** and **hydrogen carbonates**. Table 5.2 summarises the reactants and lists some common examples of each.

TRANSFER: USING TECHNIQUES TO IMPROVE MEMORY

For any topic of study, there are certain important facts. Once you know these facts, you can use them in a wide range of applications. For example, in this unit it is important that you know the chemical formulae of some common reactants and products. This includes substances such as carbon dioxide, water, hydrogen gas, oxygen gas and the common acids and bases. Choose a method that utilises your preferred learning style to learn the names and formulas of these chemicals. Check with your teacher to see if there are other chemicals whose formulae you should also remember.

TABLE 5.2 Common reactants

Name	Definition	Common examples
Acid	A chemical that has a pH less than 7. Acids neutralise bases or alkalis. All acids have the element hydrogen in them.	Hydrochloric acid (HCl) Sulfuric acid (H_2SO_4) Nitric acid (HNO_3) Carbonic acid (H_2CO_3) Ethanoic acid (acetic acid, vinegar) (CH_3COOH)
Base	A chemical that has a pH greater than 7. Bases neutralise acids. Soluble bases are called alkalis. Bases are commonly compounds of hydroxide (OH^-) or oxide (O^{2-}) combined with a metal.	Sodium hydroxide (NaOH) Magnesium oxide (MgO) Magnesium hydroxide ($Mg(OH)_2$) Calcium oxide (CaO) Sodium oxide (Na_2O)
Metal	An element found on the left-hand side of the periodic table.	Sodium (Na) Calcium (Ca) Zinc (Zn) Magnesium (Mg)
Carbonate	A compound containing carbonate (CO_3^-), usually combined with a metal.	Sodium carbonate (Na_2CO_3) Calcium carbonate ($CaCO_3$)
Hydrogen carbonate or bicarbonate	A compound containing hydrogen carbonate (also known as bicarbonate, HCO_3^-), usually combined with a metal.	Sodium hydrogen carbonate ($NaHCO_3$)

Common acids

There are some common acids for which you should know the name and **chemical formula**. Some are acids that you may find in a cupboard at home, such as vinegar and citric acid. Others are acids you may use in the science laboratory at school, such as hydrochloric acid or sulfuric acid.

TABLE 5.3 Some common acids

Name	Where it is found	Chemical formula
Hydrochloric acid	Stomach acid	HCl
Nitric acid	Some fertilisers, metal cleaners	HNO_3
Sulfuric acid	Car batteries	H_2SO_4
Carbonic acid	Soft drinks, rain, blood	H_2CO_3
Ethanoic acid (acetic acid)	Vinegar	CH_3COOH
Tartaric acid	Grapes, bananas	$C_4H_6O_6$
Citric acid	Oranges, lemons	$C_6H_8O_7$

FIGURE 5.16 Can you identify the acids found in each of these?

Reactions with metals

You have already learnt about some of the reactions of metals in Unit 4. In this unit, you will look more specifically at the reactants, products and chemical equations.

Metals reacting with acids

When some metals are added to an acid, they react to produce a salt and hydrogen gas. We can represent this reaction with the general equation:

metal + acid → salt + hydrogen gas

The salt is the compound produced from the metal and what is left of the acid after it has lost its hydrogen(s). In most cases, the salt is in solution. You are able to observe this reaction because the metal dissolves, bubbles of gas are produced and the solution warms up.

When magnesium metal is added to hydrochloric acid, a salt (magnesium chloride) and hydrogen gas are produced. This can be represented by the following equations.

Word equation:

magnesium + hydrochloric acid → magnesium chloride + hydrogen gas

Chemical equation:

$Mg(s) + 2HCl(aq) \rightarrow MgCl_2(aq) + H_2(g)$

Artists utilise the reaction between a metal and acid in etching. A metal plate is covered with a layer that will not react with the acid. The artist then makes a design by scratching into the protective layer, exposing the metal underneath. The plate is then submerged in the acid. Where the metal is exposed, it will react with the acid to leave an indentation in the metal surface (Figure 5.18).

FIGURE 5.17 Magnesium reacts with hydrochloric acid. The white colour is due to the many bubbles of hydrogen gas.

Metals reacting with water

You have previously learnt that some metals will react with water. Group 1 metals will react with water at room temperature. Group 2 metals, except beryllium (Be), will also react with water but the water may need to be heated.

FIGURE 5.18 An antique metal plate made by etching steel

FIGURE 5.19 Lithium reacting with water

These metals react with the water to produce a metal hydroxide, such as sodium hydroxide (NaOH), and hydrogen gas (H_2). The gas can be seen as bubbles of a colourless, odourless gas. The metal hydroxide is a base. Therefore, if you were to add some **indicator** to the solution, it would show a **pH** of greater than 7.

Potassium is a group 1 metal. When it is added to water, potassium hydroxide and hydrogen gas are produced. This reaction also releases a lot of energy, and can even lead to an explosion. We can write this as an equation.

Word equation:

potassium + water → potassium hydroxide + hydrogen gas

Chemical equation:

$$2K(s) + 2H_2O(l) \rightarrow 2KOH(aq) + H_2(g)$$

Acid and base reactions

When an acid and a base are combined, the acid and base are **neutralised**. In *Science 1 for the international student* Unit 4, you learnt that an acid has a pH less than 7, whereas a base has a pH greater than 7. A pH of 7 indicates that the substance is neutral. In a neutralisation reaction, the final products – a salt and water – are neutral. The salt is made up of the metal from the base and the part of the acid left after the hydrogen has been removed.

We can summarise the reaction between an acid and a base as:

acid + base → salt + water

A common acid-base reaction that you may have already performed is between hydrochloric acid and sodium hydroxide. The salt produced is sodium chloride (NaCl). This reaction can be represented by the following equations.

FIGURE 5.20 Antacids are used to neutralise stomach acid.

Word equation:

hydrochloric acid + sodium hydroxide → sodium chloride + water

Chemical equation:

$$HCl(aq) + NaOH(aq) \rightarrow NaCl(aq) + H_2O(l)$$

An indicator may be used to show when the solutions are neutralised. The colour of the indicator will depend on which one you are using. If it is an indicator with only two colours, the point of neutralisation will be shown when there is a change in colour.

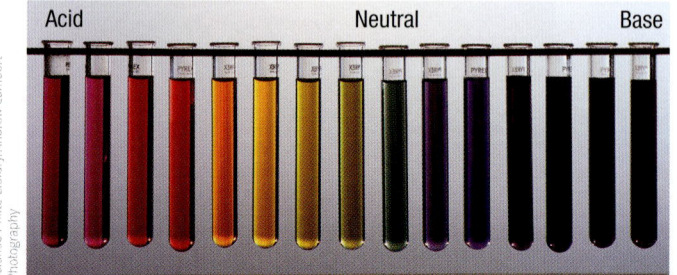

FIGURE 5.21 The colour of universal indicator in acids, bases and a neutral solution

Antacid tablets are used to treat indigestion, heartburn or acid reflux. The stomach contains hydrochloric acid, which is important in the process of digesting food. However, if this acid moves up into the oesophagus it will cause damage, resulting in the pain associated with indigestion. Antacid tablets work by neutralising the acid, thereby reducing the acidity of the fluid. There are a variety of types of antacid tablets on the market. One common brand, Mylanta, uses the active ingredients aluminium hydroxide ($Al(OH)_3$) and magnesium hydroxide ($Mg(OH)_2$).

EXPERIMENT 5.4 — Neutralisation reaction

AIM
To neutralise an acid by adding a base.

MATERIALS
Your teacher will provide you with:
- 10 cm^3 of 0.5 mol/dm^3 sodium hydroxide (NaOH)
- 10 cm^3 of 0.5 mol/dm^3 hydrochloric acid (HCl)
- **universal indicator** and colour chart
- plastic pipette or dropper
- two 10 cm^3 measuring cylinder
- four test tubes

HYPOTHESIS
Suggest and explain a hypothesis for how the pH is likely to change during this experiment.

PROCEDURE
1. Place about 1 cm^3 of the sodium hydroxide into a test tube. Label the test tube to identify what is in it. Add a couple of drops of universal indicator. What colour is the indicator? Use the colour chart to identify the pH of the solution. Record the colour and pH of the solution.
2. Repeat step 1 for the hydrochloric acid.
3. Measure 2 cm^3 of hydrochloric acid, and pour this into a clean test tube. Add a couple of drops of universal indicator to the solution.
4. Measure 4 cm^3 of sodium hydroxide. Use the plastic pipette to carefully add sodium hydroxide from the measuring cylinder to the acid solution, one drop at a time. When the solution turns green, you have completed the neutralisation reaction. What volume of base did you need to add?
5. Repeat steps 3 and 4, using 2 cm^3 of sodium hydroxide and adding hydrochloric acid drop by drop.

RESULTS
Record your data in a suitable table.

DISCUSSION
1. What do you think would happen if you added too much base in step 4? Test this to see if your prediction was correct.
2. How could you make the solution neutral again? Justify your answer. Test this to see if your prediction was correct.
3. What do you think would happen if you added too much acid in step 5? Test this to see if your prediction was correct.
4. How could you make the solution neutral again? Justify your answer. Test this to see if your prediction was correct.
5. Predict the volume of sodium hydroxide that would be required to neutralise 3 cm^3 of hydrochloric acid. Test this to see if your prediction was correct.
6. Suggest how the volumes of the solutions required for neutralisation are related.
7. Outline how well the method allowed the neutralisation reaction to be investigated.
8. Outline how the method could be improved.

CONCLUSION
What conclusions can you make about the reaction between sodium hydroxide and hydrochloric acid? Were your predictions correct?

CRITICAL THINKING: ANALYSING AND EVALUATING DATA

An important part of science is collecting and analysing data. A well-planned investigation facilitates the collection of data related to the research question. When this data is analysed, it is possible to gain a better understanding of the concept. This is how science has been able to improve our understanding of the world around us.

Acid and carbonate reaction

Another reaction that will neutralise an acid is the reaction of an acid with a carbonate. This reaction will produce a salt, water and carbon dioxide.

We can summarise this reaction by the general equation:

acid + carbonate → salt + water + carbon dioxide

One example of this occurs when acid rain falls on a limestone building. Limestone is calcium carbonate. The sulfuric acid in acid rain will react with the limestone, causing the limestone to gradually disintegrate.

Word equation:

calcium carbonate + sulfuric acid → calcium sulfate + carbon dioxide + water

Chemical equation:

$$CaCO_3(s) + H_2SO_4(aq) \rightarrow CaSO_4(aq) + CO_2(g) + H_2O(l)$$

This type of reaction can be observed by the production of carbon dioxide, which may be seen as bubbles of a colourless, odourless gas. If an indicator is added, then it is also possible to observe a colour change as the acid is neutralised.

FIGURE 5.22 Limestone contains calcium carbonate.

REVIEW

1. Classify each of the following chemicals as acid, base, carbonate, salt or metal.
   a. Lead
   b. Sodium nitrate
   c. Potassium hydroxide
   d. Sodium carbonate
   e. Sulfuric acid
2. Complete each of the following general equations.
   a. Acid + base → _____ + water
   b. Acid + _____ → salt + carbon dioxide + _____
   c. _____ + metal → salt + hydrogen gas
3. List the products formed from each pair of reactants.
   a. Zinc + hydrochloric acid
   b. Sodium carbonate + sulfuric acid
   c. Barium + water
   d. Nitric acid + magnesium hydroxide
4. What might you observe during a reaction between a piece of zinc metal and sulfuric acid?

UNIT QUESTIONS

CRITERION A

EXPLAINING SCIENTIFIC KNOWLEDGE

1. Classify each of the following reactions as either a physical change or a chemical change. (Level 1–2)
   a. Melting ice
   b. Baking a cake
   c. Mixing sand and gravel
   d. An iron nail rusting

2. Define: (Level 3–4)
   a. chemical reaction
   b. salt
   c. neutralisation.

3. List three observations that would indicate a chemical reaction has occurred. (Level 3–4)

4. Explain the law of conservation of mass, and how it applies to chemical (or symbol) equations. (Level 5–8)

5. Explain the difference between a skeletal equation and a balanced equation. (Level 5–8)

6. Explain the difference between a synthesis reaction and a decomposition reaction. (Level 5–8)

APPLYING SCIENTIFIC KNOWLEDGE AND UNDERSTANDING TO SOLVE A PROBLEM

7. Balance each of the following equations. (Level 1–6)
   a. $Mg + HCl \rightarrow MgCl_2 + H_2$
   b. $S_8 + O_2 \rightarrow SO_2$
   c. $Al + Br_2 \rightarrow AlBr_3$

8. A chemical reaction releases a colourless, odourless gas. Explain what you would do to prove that the gas is hydrogen gas. (Level 1–6)

9. During the Haber process, nitrogen gas (N_2) and hydrogen gas (H_2) react to produce ammonia (NH_3). (Level 7–8)
   a. Write a word equation to represent the reaction.
   b. Write a balanced chemical (or symbol) equation for the reaction.
   c. What type of reaction occurs during the Haber process?

10. When solid sodium hydrogen carbonate ($NaHCO_3$) is heated, it forms solid sodium carbonate (Na_2CO_3), carbon dioxide gas (CO_2) and water (H_2O). (Level 3–6)
    a. What type of reaction is taking place?
    b. Write a word equation to show what occurred during the reaction.
    c. Write a balanced chemical (or symbol) equation for the reaction.

11. Predict the products of the reaction between iron and hydrochloric acid. (Level 3–6)

12. Many camping stoves use the combustion reaction of butane (C_4H_{10}). Describe why the instructions say to use the stove in a well-ventilated area. (Level 3–6)

13. A student mixed up the labels for three colourless solutions. She knows that they are sodium carbonate, sodium hydroxide and sodium chloride. Explain what tests you could do to determine which solution is which. (Level 7–8)

14. Explain why it is important to consider the consequences of combining chemicals. (Level 1–6)

INTERPRETING INFORMATION

15. When acid rain containing sulfuric acid falls on limestone buildings, which contain calcium carbonate ($CaCO_3$), the buildings gradually erode. Explain this observation using your knowledge of chemical reactions. (Level 1–4)

16. Use the following reaction to answer the following questions. (Level 1–4)
    $MgCO_3(s) + 2HCl(aq) \rightarrow MgCl_2(aq) + H_2O(l) + CO_2(g)$
    a. State the type of reactants in this reaction.
    b. Describe the observations that you would make during the reaction.
    c. Describe what you would do to identify the gas produced.

17. Food is often stored in cans with a layer of tin (Sn) on the inside; this tin resists corrosion. Acidic foods may cause the layer to react and lead to health problems if the food is consumed. What might you observe when you open a can that has reacted with its contents? Use a general equation to support your prediction. (Level 5–8)

18. Identify the type of chemical reaction represented in Figure 5.23. Justify your answer. (Level 5–8)

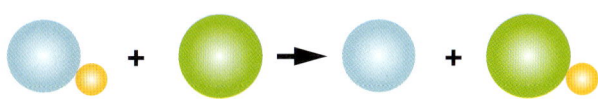

FIGURE 5.23

REFLECTION

1. The key concept during this unit was change: chemical change. Compare how we used the concept of change in this unit with how the concept of change is used in other subjects.
2. We considered chemical systems in this unit. Explain what the term 'systems' means in this context.
3. You have seen how chemists use models of atoms and molecules to help us understand chemical reactions. Explain why models are useful in science.
4. Do you think we can explain all chemical reactions? For instance, look at the reactivity series, and predict the reaction of aluminium with acids. Find out if your prediction is correct.
5. You will have noticed over this unit how the conditions under which a chemical reaction takes place can affect the reaction. Describe some examples of this.

UNIT 6
WAVES: LIGHT AND SOUND

KEY CONCEPT
Communication

RELATED CONCEPTS
Energy
Development
Consequences

GLOBAL CONTEXT
Scientific and technical innovation – an exploration into how our knowledge of waves is leading to innovative communication technologies

STATEMENT OF INQUIRY
Developments in technology are expanding our forms of communication, often with unknown consequences on our own lives.

INQUIRY QUESTIONS

FACTUAL
1. What are the characteristics of waves?
2. What are the differences between sound and light waves?
3. How do we see and hear?

CONCEPTUAL
4. How do we calculate frequency and wavelength?
5. How do we help people who have hearing and sight disabilities to communicate?
6. How does a change in energy influence a wave?

DEBATABLE
7. Is digital sound better than sound from analogue devices?
8. How is our use of modern communication technology affecting our lives?

Introduction

Waves are all around us. Human bodies have evolved to use light waves to see and to use sound waves to communicate. We use different types of waves in hundreds of ways: to diagnose and treat ourselves for medical conditions, to cook, to check for forgeries, to investigate crime, to phone

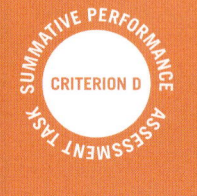

Go to http://mypsci3.nelsonnet.com.au and click on **Communication technology study** to read the findings of this study.

AFFECTIVE: RESILIENCE

This is a challenging assessment task. Consider how well you cope with the challenge. How well do you cope with difficulties? Do you stay positive and show perseverance?

Considering the impact modern communication technology is having on our lives

The Engineering Design Centre at the University of Cambridge in the UK recently carried out a study into the impact of modern communication technology on our lives. Their research brief was written as:

> Modern communication technology makes it possible to stay connected anywhere, all the time, and the flow of information is nearly limitless. With all the benefits afforded by this newfound capability, however, come potential consequences. Following the ever increasing flow of information through our computers, televisions, and phones has been a stream of concerns about the change in how we, as humans, communicate. Will the new ways in which we acquire, process, and relate information in turn change us as individuals, families, and societies?

'Culture, Communication and Change: An investigation of the use and impact of modern media and technology in our lives', © 2015 Cambridge EDC

Your challenge

Imagine you work in a research team at a university and would like to carry out similar research. Design a questionnaire, present your results as a graph, and make conclusions and suggestions based on your data. Work in small groups on one of the areas mentioned above, then share your findings during a class conference.

This might help

Focus your research around the following areas.
1. The kind of communication technology people are using. What are the most popular communication technologies? Which are people using? How many hours do they use it a day? When do they use it?
2. Attitudes to communication technology. Do people feel in control of its use? Do they think it has too much impact on their lives? Are they worried about safety? Have they suffered cyber bullying? Do they feel any form of addiction? Are they considering reducing their use?
3. Impact of communication technology on their lives. Is it having any positive or negative impact on their relationships with family or friends? What do they see as the main advantages offered by modern communication technology? Do they prefer normal face-to-face communication or the use of technology?
4. The need for guidelines for the use of communication technology. Are people happy with the location of the communication technology in their homes? Would it be better to have all computers in a common space (and not in bedrooms)? Do they accept that rules are needed, such as no use at the dinner table and/or limitations on use during homework, numbers of hours online, numbers of messages sent, or sites that can be visited?

Consider the types of communication technology to include in the study – email, instant texting (SMS and Whatsapp), voice via telephone, video calls (such as Skype), social networking, gaming, Instagram and/or any others you feel are suitable.

Your questionnaires could: (i) ask people to write written answers, (ii) be answered as yes or no or (iii) be answered on a scale such as never, rarely, regularly, all the time. The use of scales of this last kind is likely to provide the best data.

each other, to broadcast television and so on. Some of the technologies we have developed to make use of waves have had a huge effect on our society, and new technologies continue to emerge and impact the way we live and communicate. Modern communication technologies such as texting, face-to-face video, social media, emails, voice to voice, gaming and others are transforming the way we interact with our friends, families and as a society.

What are waves?

Perhaps your first thoughts about waves are surfing waves and the wave of the Queen of England. However, you also know about water waves and sound waves. Although they are very different, they have one thing in common – energy. To create a water wave, you supply energy by dropping a stone into a pond or dipping your hand in and out of a bowl of water. To create sound waves you can use your **vocal cords** to speak or you can transfer some energy to a desk by tapping it with your ruler.

FIGURE 6.1 Ripples on a pond

Waves are a form of energy transfer. Waves are **vibrations** that travel outwards, away from where they start. Try putting your fingers gently on your throat when you make a sound or putting your fingers on the desk a small distance from where you are tapping it. Both waves travel through substances, but don't actually carry the substance along with them. People often believe that water waves carry water along with them, but if you look carefully at some water as waves travel through it, you will see that the water stays in the same place. Try putting a cork on the water. The waves move through the water and make the cork bob up and down, but they don't move the water or the cork from one place to another. The cork doesn't travel sideways, just up and down.

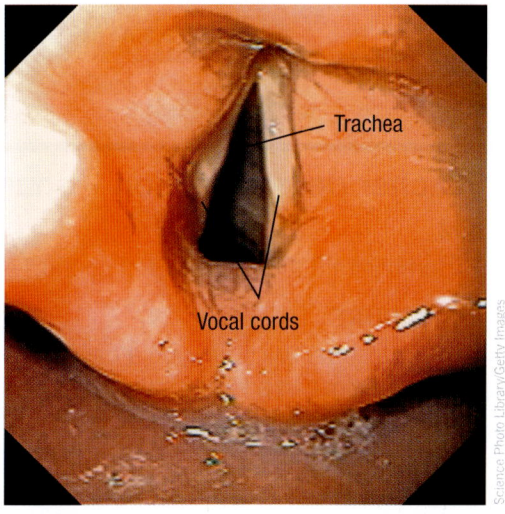

FIGURE 6.2 A close-up image of human vocal cords

Waves transfer energy from one place to another but don't transfer the substance they travel through with them, as you can see in Figure 6.3. A cork moves up and down, but stays in the same place in the water; in other words, it does not move from side to side. Water is the **medium** (plural: **media**) that allows the energy to move through. Most waves need a medium to travel through.

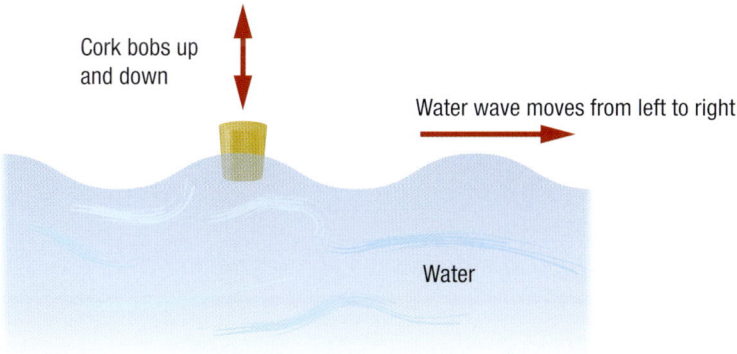

FIGURE 6.3 Arrows show the direction of the wave (left to right) and the direction of motion of the cork (up and down).

Sound waves

Air is a very important medium for sound waves. Air consists of many particles that are moving around in **random** directions. When an object such as a loudspeaker emits a sound, it vibrates. The vibrations are **transmitted** to the air particles, making them vibrate too. These vibrations are passed from air particle to air particle as they bump into each other.

FIGURE 6.4 Sound travels through a medium.

Go to http://mypsci3.nelsonnet.com.au and click on **Wave on a string** to see how the energy is transferred from particle to particle.

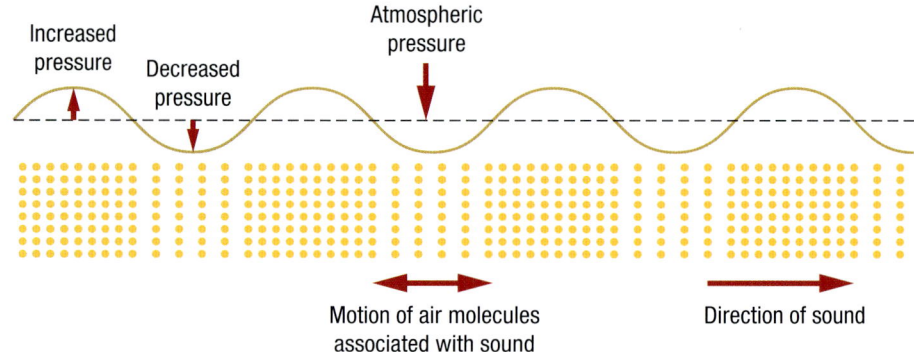

FIGURE 6.5 Air particles vibrate backwards and forwards in the same direction as the direction of the sound wave.

Teacher demonstration – sound

EXPERIMENT 6.1

AIM
To demonstrate the properties of sound.

PROCEDURE
1. Your teacher will show you a candle flame placed close to a loudspeaker as it makes some different sounds. What do you observe? Can you use ideas about particles to explain your observations?
2. Your teacher will place a ringing bell into a special jar called a bell jar. Observe what happens when the air in the jar is removed to create a vacuum. Try to use ideas about particles to explain your observations. Can sound travel in a vacuum?

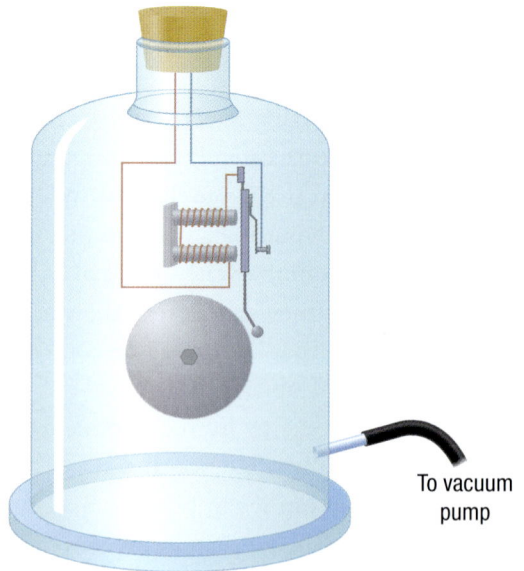

FIGURE 6.6 The bell jar experiment

How does the ear detect sound waves?

You hear sound waves because moving particles vibrate and these vibrations help you hear. A sound is emitted by means of a disturbance (i.e. energy) to the medium (air). The air particles vibrate and the vibrations reach the ear. From the outer ear, the vibrations travel down the ear canal and reach the eardrum, causing it to vibrate as well. The vibrating eardrum causes the ossicles (three tiny bones called the hammer, stapes and incus) in your ear to vibrate. These vibrations get sent to the cochlea, which is filled with liquid and tiny hair cells. The movement in the fluid, caused by the oscillating bones, causes the hair cells to send messages to a nerve connected to the brain, which interprets the signals as audible sound. Our brain then gives us information about the **volume**, the **pitch** and the possible location and source of that sound.

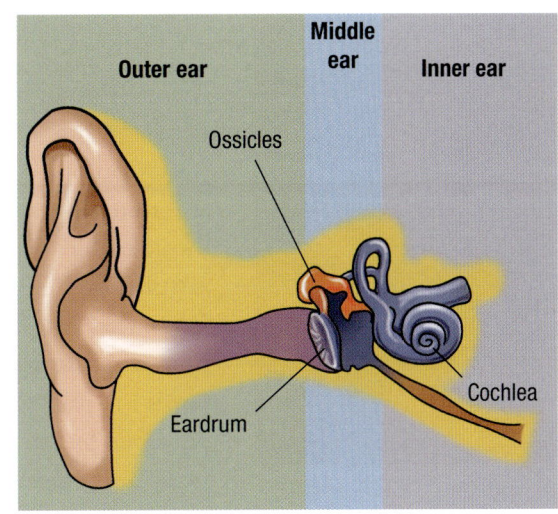

FIGURE 6.7 The structure of the human ear

ACTIVITY — Helping people with hearing disabilities

The hearing process involves many factors and sometimes things go wrong. For example, some people are born without liquid in their cochlea, making it impossible for them to hear. Your task is to research the different hearing disabilities that exist and to choose one specific problem to research further.

Create a display, using a technology of your choice, that demonstrates how science is addressing this problem (for example, the cochlear implant in Figure 6.8). Include a discussion about a factor that can restrict the success of the scientific solution. Use good scientific vocabulary throughout and reference all the sources that you have consulted.

TA SUPPORTING DEAF PEOPLE

After you've completed the activity, you might realise how lucky people are who have good hearing. In some countries, people who are part of the deaf community face discrimination and do not have access to equal opportunities, such as education or employment. However, there are several charities that help deaf people by providing access to opportunities such as training. Consider some creative ways you and your classmates can raise both awareness and money to help deaf people in these countries. If you live in a country that discriminates, invite someone from the deaf community into your school to speak with the students, and find ways to reach out and help people in the deaf community.

Go to http://mypsci3.nelsonnet.com.au and click on **Deaf child worldwide** for information on a charity committed to supporting deaf children in developing countries.

Louder than loud

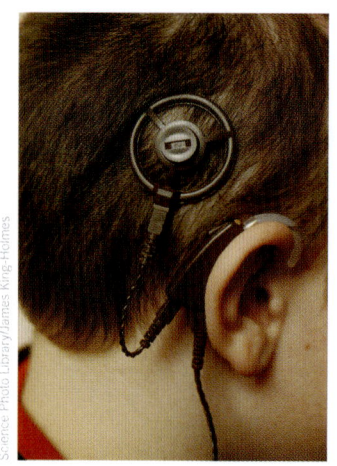

FIGURE 6.8 A cochlear implant

One way you can measure a sound's volume is by using an **oscilloscope**. An oscilloscope is a device that can produce a waveform of a sound. The height of the waveform changes according to the volume of the sound. We do not usually measure the actual height of the waveform. Instead we measure its **amplitude**. The amplitude of a wave is half the total wave height. Consider Figure 6.9 – the white line in the middle is the **rest position**, also known as the **equilibrium position** – that means this is the place where the particle would have been if it was not disturbed by the energy. The very top bit of the wave is called the **peak** or the **crest**, while the very bottom bit is called a **trough**. The distance covered from peak to peak (or from trough to trough) is the **wavelength**.

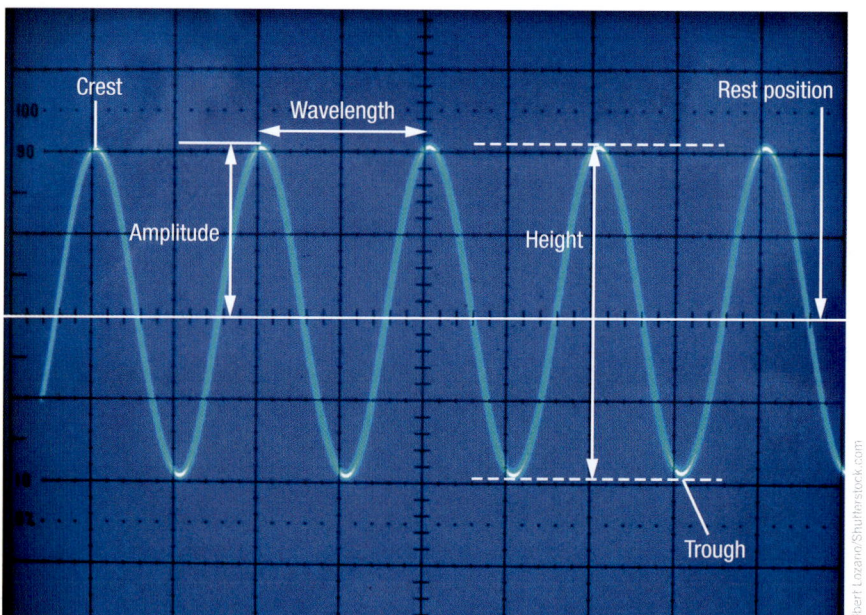

FIGURE 6.9 A sound wave graph

Using an oscilloscope to show volume of sound waves

EXPERIMENT 6.2

Your teacher will demonstrate some sounds of different volumes shown on an oscilloscope screen. Make notes of your observations.

What is the relationship between the amplitude and volume of the sound?

REVIEW

1. List five examples of waves.
2. Describe what is meant by a 'wave'.
3. The tagline for the film *Alien* (1979) was, 'In space no one can hear you scream'. Explain whether or not that tagline was scientifically correct.
4. Are the following statements true or false? Rewrite the false statements to make them true.
   a Sound waves cannot travel through the medium of water.
   b Sound waves can travel through concrete.
   c Sound waves are transmitted by particle vibrations.
   d As sound waves travel through a medium, particles do not lose energy when they bump into each other.
   e Sound waves enter the ear through the cochlea.
5. Describe how the sound wave vibrations in a human ear are transmitted from the eardrum to the cochlea.
6. Describe one cause of a hearing problem and how it can be avoided.

Pitch and frequency

Whenever we hear a song, we can hear different notes being sung and we can also hear that some notes are higher than other notes. High notes and low notes have different pitches. Pitch depends on how quickly the particles of a medium vibrate. The more vibrations there are every second, the higher the pitch. The number of vibrations per second is called the **frequency** (because it is a measure of how frequently the wave vibrates). The frequency of a wave is measured in **hertz** (Hz), where $1\,\text{Hz} = 1\,\text{s}^{-1}$. Middle C note on a piano has a frequency of 262 Hz. That means that when a musician plays that note, the string inside the piano vibrates 262 times per second. The greater the frequency, the higher the pitch.

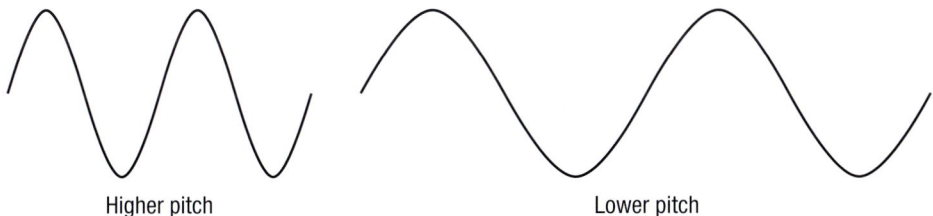

Higher pitch   Lower pitch

FIGURE 6.10 These two waves have the same amplitude but different frequencies.

Humans can hear sounds in the frequency range of 20–20 000 Hz, but dolphins have a much bigger range of hearing. There is evidence that they can make and hear sounds of frequencies in the range of 0.5–200 000 Hz.

INVESTIGATION 6.1

Pitching it right

YOUR CHALLENGE
To investigate the factors that affect the pitch of a vibrating object.

THIS MIGHT HELP
You don't necessarily need musical instruments to make differently pitched sounds. Tuning forks, objects that make a sound when tapped gently, and water-filled test tubes can all be used to make sounds of different pitches.

Carry out and write up your investigation following the guide in Appendix 3 on page 177 or as advised by your teacher.

EXTENSION
Research the frequencies that different animals and people (including deaf people) can hear. Try to recreate the appropriate pitches to design a music instrument that is easy for hard of hearing and deaf people to hear.

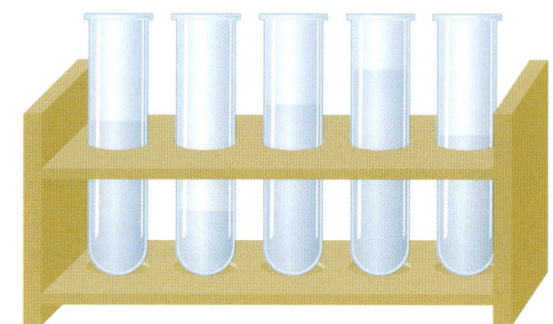

FIGURE 6.11 Test-tube musical instrument

ACTIVITY

Measuring the sound profile of your classroom

The level of sound in a classroom has a proven effect on students' concentration, mood and learning. The volume (level) of sound is normally measured in decibels (dB). A commonly used value for recommended maximum sound levels in school classrooms is 35 dB. Your teacher will use a sound meter and data logger to construct a sound profile of your classroom over time. Reflect on your experience of hearing the sounds and considering the data.

TA THINKING ABOUT NOISE

While you were reflecting on the sounds in your classroom, some of you might have thought that it was very noisy and other students may not have been bothered by the noise. What you love to hear might not sound quite so good to your friends or your teachers. In your house, your parents might ask you to keep the music level down when it is not noisy for you. Discuss in small groups some sounds that you do and don't like to hear. Can noise cause stress? Specifically think about background noise for people wearing hearing aids or cochlear implants. Share your ideas as a class.

How far can sound go?

Sound waves travel as vibrations of particles. As each particle passes on the vibration to the next, some energy is lost, mainly as a result of friction. As the sound wave travels further and further from its source, it spreads out in all directions. It loses more and more energy until eventually the vibrations are so weak that you can no longer hear the sound. Therefore, the distance sound can travel depends on the initial energy as well as the friction between the particles.

FIGURE 6.12 Sound can travel long distances.

ACTIVITY Experiences of deaf people

Read the articles in the weblink. After reading these articles, you will have a small insight into the life of someone with a hearing disability. Summarise the information and come up with some ideas to better include people with a hearing disability in the classroom.

Go to http://mypsci3.nelsonnet.com.au and click on **Experiences of deaf people** to read these articles.

INVESTIGATION 6.2 Sound travelling through materials

YOUR CHALLENGE
To investigate some variables that affect how sound travels through materials.

THIS MIGHT HELP
This is a very 'open-ended' investigation. You will need to define carefully which variables you will test. Some obvious independent variables include the type of material, or the tension in the material if it is a type of wire or string. What will be the dependent variable? For instance, it could be the loudness of the sound received, or the speed of the sound through the material.
   Carry out and write up the investigation following the guide in Appendix 3 on page 177 or as advised by your teacher.

SAFETY
Check with your teacher before you start transmitting any sounds or choosing materials and shaping them.

COMMUNICATION
Consider how best to design a results table to communicate experimental results.

The speed of sound

During a storm, we are used to seeing a flash of lightning and then hearing a crack of thunder. The thunder and lightning happen at the same place at exactly the same moment, but we see the light before we hear the sound. You might have been told that by counting the time lapse between them, you can work out your distance from the storm.

Two sprinters finish a race at different times because they run at different speeds – one runs faster than the other. The same is true for sound waves and light waves.

Light travels faster than sound. Light waves travel at 300 million metres per second in air. To find out the speed of sound, we need to know the time it takes for it to travel a measured distance.

FIGURE 6.13 A lightning strike

Sound as a form of long-distance communication

A number of cultures use drumming to transmit messages. These include the Tama in Senegal, and the Doodo in Mali. This has been referred to as a drumming language, or talking drums. The talking drumming mimics real language and can convey messages up to 11 km. The practice spread to the USA and the Caribbean with the slave trade. Slavers tried to ban the use of talking drums because it was such a strong cultural force and was used for communication. Today, modern music often incorporates talking drums, especially in West Africa. It is considered to be a vital type of music, which plugs directly into people's emotions.

FIGURE 6.14 Dundun drums from Mali

EXPERIMENT 6.3 — The speed of sound

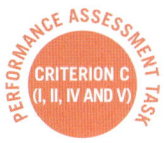

AIM
To calculate the speed of sound in air.

MATERIALS
- two wooden blocks
- stopwatch
- trundle wheel or very long tape measure

PROCEDURE
This experiment relies on finding a space in the school grounds where sound can reflect from a solid surface such as a wall or the side of a building. You need a good **echo.**

1. With a partner, find a space outside to bang together the wooden blocks to make a sharp, clear 'clap'. You should do this a good distance (more than 100 metres if possible) from a wall or similar surface so that you can hear the echo from your clapping sound.
2. Measure the shortest distance from your position to the wall.
3. Clap the wooden blocks together and listen for the echo.
4. Repeat step 3 and use the stopwatch to measure the time between the clap and its echo. This will be difficult to do accurately because it is a very short time.
5. Try to keep clapping the blocks together at exactly the same time as you hear the echo, so that you set up a clapping rhythm. Measure the time for 10 claps.
6. Repeat step 5 at least two more times.

RESULTS AND CALCULATIONS
1. Record your distance and time results in a suitable table.
2. Divide the average time by 10 to find the time for the sound to travel to the wall and back again.
3. Multiply the distance to the wall by two, to find the total distance the sound travelled.
4. Use the following equation to calculate the speed of sound:

$$\text{speed (m/s)} = \frac{\text{distance (m)}}{\text{time (s)}}$$

CONCLUSION
Compare your value for the speed of sound in air to the average value of 340 m/s given in data books. Compare the speed of sound with the speed of light and comment on the difference.

EVALUATION AND DISCUSSION
How accurate do you think this method is for finding the speed of sound in air? What problems did you find? Suggest possible ways to reduce or solve the problems with the experiment.

EXTENSION
Many of us have been taught to count the seconds, '1 elephant, 2 elephant, 3 elephant . . .' or something similar, between seeing a lightning flash and hearing the thunder. Every 5 seconds we count tells us the storm is one mile further away. One mile is approximately 1600 metres. Considering the results of your experiment and the difference between the speed of light waves and sound waves, show whether this method can work or not.

> **REVIEW**
>
> 1. Describe what is meant by the word 'frequency'.
> 2. What are the units of frequency?
> 3. What is the frequency of middle C on a piano?
> 4. Predict which of the following will result in a sound of higher pitch: a long tuning fork or a short tuning fork.
> 5. Write definitions that describe the different meanings of the words 'sound' and 'noise'.
> 6. Give an example of one factor that can affect the distance that a sound can travel from its source.
> 7. A student sees a flash of lightning and counts 8 seconds until she hears the thunder. Calculate how far away from her the lightning flash happened.
> 8. Describe a piece of evidence (other than thunder and lightning) to demonstrate that the speed of light waves is faster than the speed of sound waves.

Light

Light is a wave too. It transfers energy from one place to another without transferring the medium. Light does not require a medium. Light belongs to a family of waves called the **electromagnetic spectrum**. These waves are not transmitted by vibrating particles, but by vibrating electric and magnetic fields. You will learn more about the electromagnetic spectrum later in the unit. First, we will investigate how light behaves.

FIGURE 6.15 The non-luminous Moon eclipses the luminous Sun.

Light and sight

Just as you need sound waves to hear, you need light waves to see. Light waves must come from an object and enter your light-detection devices – your eyes. Your eyes detect the light and convert it to an electrical signal that is transmitted by nerves to your brain. The brain then gives you information about the source of the light wave. For us to see an object, either the object itself can produce the light wave that enters our eyes, or the light wave can be **reflected** from it. Objects that produce their own light are **luminous**. Objects that do not are called **non-luminous**. Try to identify as many luminous and non-luminous objects as you can.

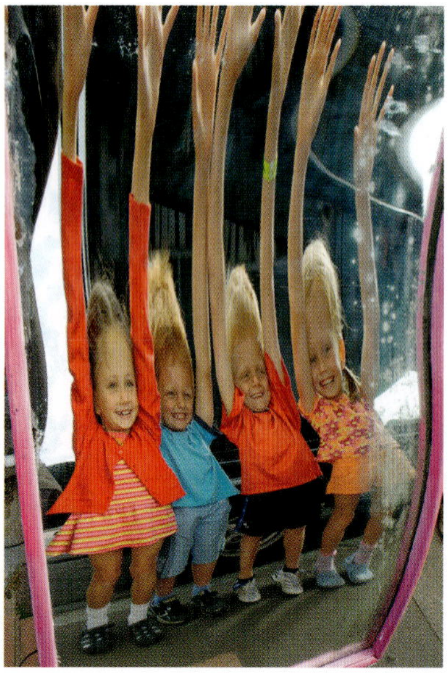

Mirror, mirror

To see ourselves, we look in the mirror. Light from objects such as ourselves is reflected from the mirror. Light can only travel in straight lines between objects. You can probably think of evidence that demonstrates that light travels in straight lines.

FIGURE 6.16 Fairground mirrors can distort your reflection.

EXPERIMENT 6.4 Mirror image

AIM
To investigate the image formed by a **plane (flat) mirror**.

MATERIALS
- plane mirror
- book with writing
- yourself

PROCEDURE
1. Look at yourself in the mirror.
2. Change the distance between the object (yourself) and the mirror and record your observations about the image of yourself and how it changes with distance.
3. Change the angle between the object and the mirror and record your observations.
4. Look at some writing in the mirror and record your observations about the differences between the object (the writing when you look at it directly) and the image (the writing when you look at it in the mirror).

RESULTS AND CALCULATIONS
Record your observations in a suitable scientific style.

CONCLUSION
What have you found out from your observations? Try to suggest how the light waves travel from the object to your eyes. Draw a diagram. Use straight lines with arrows to represent the direction of the light waves.

EVALUATION AND DISCUSSION
Discuss with a partner where you think the mirror image is formed and why it is back to front. Discuss your findings as a class.

EXTENSION
Write your name or a message so that it appears the correct way round in a mirror. Try the experiment again using curved mirrors instead of plane ones.

Mirrors have other uses apart from checking our appearance. Drivers need them; so do dentists. Mirrors can send light in new directions. Imagine a world without mirrors. Would it be a better place or a more dangerous one?

Heliographs are a form of communication that uses mirrors. They are based on sending flashes of light from the Sun by pivoting a mirror in the appropriate direction. They were often used to send Morse code. Heliographs were used a lot during the late 19th and early 20th centuries and were used until recently by the British and Pakistani armies. The Greeks used to send messages by reflecting flashes of light from their shields in 400 BCE.

The Hubble Space Telescope has a mirror with a diameter of 2.4 metres and a mass of 848 kg. The original mirror had slight imperfections and had be repaired to allow it send back high-quality images to Earth.

Go to http://mypsci3.nelsonnet.com.au and click on **Reflection and mirrors** to investigate ray diagrams that show us how mirrors work.

FIGURE 6.17 A heliograph for sending messages by flashing light from the Sun, around 1900

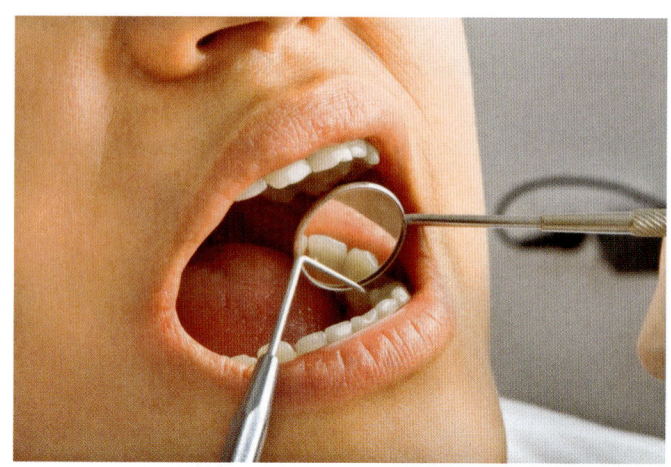

FIGURE 6.18 Mirrors are used by dentists.

How many images?

EXPERIMENT 6.5

PERFORMANCE ASSESSMENT TASK — CRITERION C (I, II, IV AND V)

ATL

CRITICAL THINKING
Consider the importance of using a suitable range of measurements when designing a scientific investigation.

AIM
Investigate the relationship between the angle of two mirrors and the number of images formed.

MATERIALS
- two small plane mirrors
- protractor
- small object such as a stone
- plasticine

PROCEDURE
1. Set up your mirrors and small object as shown in the diagram.
2. Count the number of images of the object you can see in the mirrors.
3. Change the angle of the mirrors to 90° and count the number of images again.
4. Repeat the observations for the two mirrors every 10° at angles from 80° down to 20°.

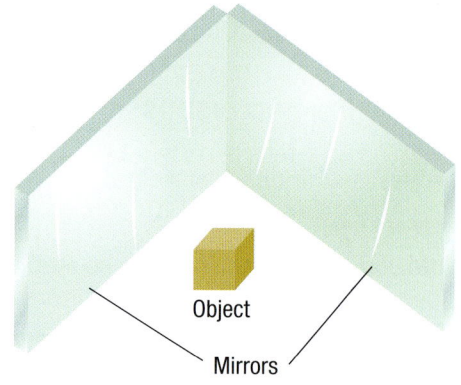

FIGURE 6.19 How many images can you form?

RESULTS AND CALCULATIONS
Record your results in a suitable table.

CONCLUSION
Comment on whether or not there is a pattern in your results. Is there a **correlation** between the angle of the mirrors and the number of images they formed?

EVALUATION AND DISCUSSION
Could the reliability, precision and validity of the results be improved for this experiment? Make suggestions for improvements.

SAFETY
If you are using glass mirrors, make sure they are fixed securely and safely so there is no danger they might fall onto your face.

> **ACTIVITY** **Kaleidoscopes**
>
> **YOU WILL NEED**
> - some small plane mirrors (preferably plastic)
> - cardboard or cardboard tubes
> - glue
> - scissors
> - colouring materials
> - tracing paper
> - some shiny materials, e.g. foil, sweet wrappers, glitter
>
> **WHAT TO DO**
> Make your own version of a **kaleidoscope**. You might need to do some research to help with your design before you get started.
>
> **WHAT DID YOU DISCOVER?**
> How many mirrors did you need and how did you angle them? Draw a diagram of your design and try to explain how the reflections were formed.

FIGURE 6.20 Inside a kaleidoscope

Periscopes

Periscopes are used for observation over and around an object or obstacle that prevents direct line-of-sight observation. They have been used in submarines since around 1900. They were sometimes fixed to rifles in World War I to help soldiers see over the top of trenches.

FIGURE 6.21 Australian soldiers using periscopes in Gallipoli in World War I

> ### Up periscope
>
> Another device that uses the reflection of light is the periscope.
>
> #### YOU WILL NEED
> - some small plane mirrors (preferably plastic)
> - cardboard or cardboard tubes
> - glue
> - scissors
>
> #### WHAT TO DO
> Make a periscope. You'll need to think about how to position the mirrors to get the reflections you need. You might need to do some research to help with your design before you get started.
>
> #### WHAT DID YOU DISCOVER?
> How easy was it to get the light to reflect to do the job you wanted? Draw a diagram to show how the light reflects in your periscope.

ACTIVITY

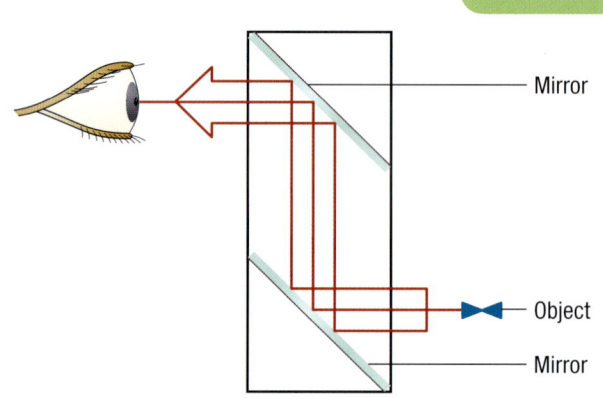

FIGURE 6.22 A simple periscope

Total internal reflection

Modern periscopes use **prisms** instead of mirrors. A prism can reflect light from its inner surface by a process called **total internal reflection**.

FIGURE 6.23 A prism

ACTIVITY: Total internal reflection

YOU WILL NEED
- two right-angled prisms
- ray box with a power supply and single slit
- plain paper

WHAT TO DO
Shine a single light ray at one prism. Rotate the prism until the light ray enters the prism, reflects off the inside surface of the prism and then leaves again at 90°. Draw this set-up on the plain paper. Carefully draw the paths of the light ray inside and outside the prism. Now use total internal reflection from the two prisms to show how they could be arranged for use in a periscope. Draw the arrangement.

WHAT DID YOU DISCOVER?
Could prisms be used in exactly the same way as mirrors to make a periscope? Are there any advantages to using prisms instead of mirrors in devices such as periscopes?

SAFETY
A ray box contains a bulb that gets very hot when it has been switched on for a while. Take care when handling the ray box.

REVIEW
1. Describe a piece of evidence to show that light cannot curve around an object.
2. What is the name of the part of the eye that contains light-sensitive cells?
3. Outline the journey of a light wave through the eye.
4. Name one common sight problem and describe how it can be solved.
5. What is the scientific name for a flat mirror?
6. Describe three everyday uses for mirrors apart from simply checking your appearance.
7. Give an example of a device other than a periscope that uses prisms to reflect light.

The electromagnetic spectrum

Recall that light waves are part of a family of waves called the electromagnetic spectrum. Just like members of any family, these waves have similarities, but also differences. They all travel at the speed of light. Unlike sound and water waves, light waves don't need to travel through a medium (made up of particles). This means they can travel through empty space. Think about our Sun. What evidence is there that light can travel through space?

All electromagnetic waves have different frequencies. This means they all carry different amounts of energy and so they can all be used for different jobs.

The members of the electromagnetic spectrum family

The electromagnetic spectrum is made up of the following waves, in order from lowest frequency (and energy) to highest: radio, microwave, infrared, visible light, ultraviolet, X-ray, gamma ray.

FIGURE 6.24 Most television remote controls operate by infrared waves.

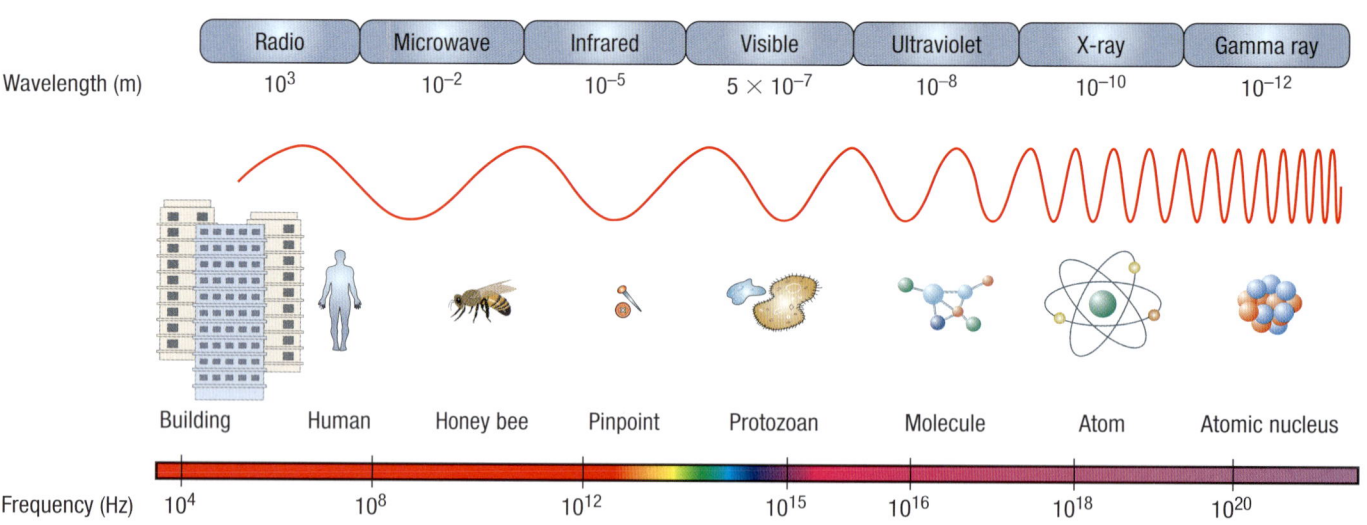

Radio waves: Broadcasting, communications and satellite transmission
Microwaves: Cooking, communications and satellite transmission
Infrared: Cooking, thermal imaging, short-range communication, optical fibres, television remote controls and security systems
Visible: Vision, photography and illumination
Ultraviolet: Security marking, fluorescent lamps, detecting forged bank notes and disinfecting water
X-rays: Observing internal structure of objects, airport security scanners and medical X-rays
Gamma rays: Sterilising food and medical equipment, and detection and treatment of cancer

FIGURE 6.25 The electromagnetic spectrum

Despite some of their names, they are all waves. They all transfer energy from one place to another without moving their medium along with them. We use light to see, but it also has other uses such as lasers.

FIGURE 6.26 We use electromagnetic waves to communicate.

ACTIVITY: Uses of electromagnetic waves

YOU WILL NEED
- large sheets of paper or mobile devices for recording information
- pens
- other display materials
- access to library, reference books, internet etc.

WHAT TO DO
In small groups, research and discuss each of the waves of the electromagnetic spectrum. What do you know about them and how they are used? What other information can you find out? Write down all your ideas and discoveries and use them to make a poster about the electromagnetic spectrum. Your poster should contain information about each type of wave in the electromagnetic spectrum and could include information about: source, use, detection, frequency and wavelength range, and possible danger of each wave type.

The waves of the electromagnetic spectrum are a hugely important part of our lives. Over the last century or so, the invention and manufacture of devices that make use of electromagnetic waves have changed the way we live. Consider how big a change they have made. Without electromagnetic waves, there would be no radiotherapy, no remote controls, no mobile phones, no televisions, no X-ray machines; the list goes on and on . . .

Modern communication technology

FIGURE 6.27 Mobile phone antennas in California

As is the focus of the summative performance assessment task for this unit, our lives are being radically transformed by access to new communication technologies. You are all digital natives who were born well after the invention of the internet, social networking and mobile phones, so for you these rapid and radical changes may not seem as dramatic as they do to your parents and grandparents.

Information is sent to and from mobile phones via microwaves, similar to the electromagnetic radiation used in microwave ovens. You are probably aware of the transmitting and receiving mobile phone antenna in your area.

The advantage of microwaves is that they have a small wavelength, which allows the use of smaller antennas that can direct the microwaves in narrow beams and without interference with other beams. Microwaves can also carry a high level of information. The disadvantage is that as their frequency is high they cannot reflect around hills or mountains, so more transfer stations are needed.

The digital information age

There are two methods of sending information – using digital signals or using analogue signals. Modern communication technology consists of digital signals. This means the information is converted into binary code. Binary code uses just two digits, 0 and 1, rather than our normal 1–10 system. The 0s and 1s are then sent 'down the line' (telephone, optic fibre, TV signals, satellite transmission, from the CD) as a series of electrical impulses or flashes of light in optic fibres). We can describe them as a series of ONs and OFFs.

Digital signals have many advantages – they are easier to process and store, and there is less distortion in the signals as they are transmitted. In theory, a digital copy of a document is always a perfect copy. Analogue signals are just changing amounts of electricity down a wire, and this is easily distorted. Ask your parents about cassette tapes and how international phone calls used to be.

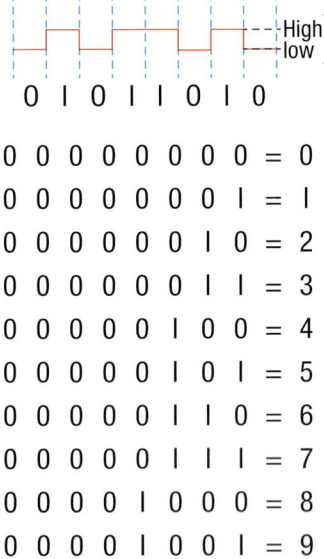

FIGURE 6.28 How 8-bit binary code can be used to produce the numbers 0 to 255

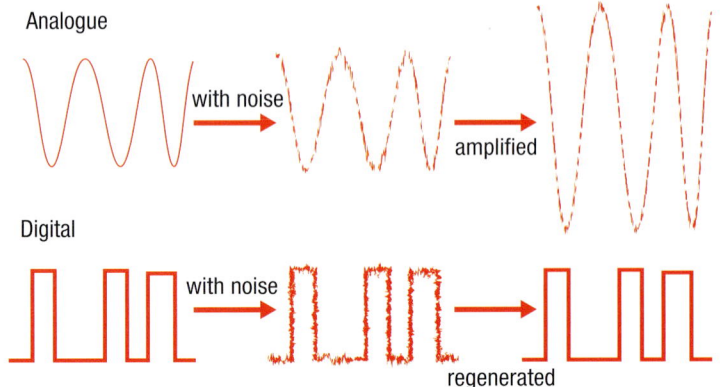

FIGURE 6.29 Digital signals transmit with less distortion than analogue signals.

Optical fibres

An increasingly common way of sending information to and from your house (TV, internet, mobile phones if using wifi) is via optical fibres. A large amount of digital information can be sent via flashes of light (on and off). In the optical fibre, light reflects inside in a process called total internal reflection. An advantage of optical fibres is that very little light is lost through the edges, so it can travel several kilometres before it needs amplification.

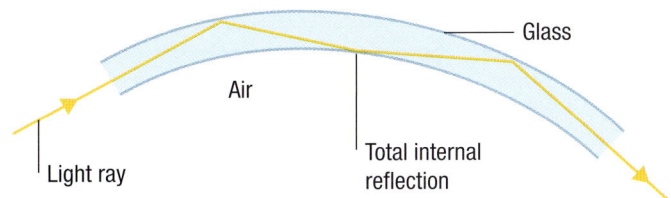

FIGURE 6.30 An optical fibre works because of total internal reflection.

Future developments in modern communication technology

The future of communication is hard to predict. The pace of development over the last 20 years exceeded everyone's expectations. People talk a lot about augmented reality, such as is being opened up by the use of Google Glass. The idea of brain–computer interfaces is being seriously considered.

Brain waves and learning

The brain also has electrical wave patterns. Most of our thinking takes place in the beta state, 12–25 Hz. One of the secrets of effective learning is to try to slow down your brain into the slower alpha and theta states. In these states, you become more relaxed and focused on your learning. You are also likely to become more creative. Being relaxed and focused is an important part of doing well in summative assessments such as tests. As a class, you could discuss the different ways students can help slow down their brain waves.

TRANSFER
Knowledge about learning, and how the brain works.

> ### REVIEW
>
> 1 Describe what is meant by the electromagnetic spectrum.
> 2 Name a common feature of the waves of the electromagnetic spectrum.
> 3 Name one source for each wave in the electromagnetic spectrum.
> 4 State which of the following is the odd one out.
>   ultraviolet, radio, visible light, gamma rays, sound, infrared
> 5 Outline why we use X-rays rather than gamma rays to check for suspected broken bones.
> 6 Outline how mobile phone signals are sent.
> 7 Describe what we mean by digital signals and explain why they are preferable to analogue signals.

UNIT QUESTIONS

CRITERION A

EXPLAINING SCIENTIFIC KNOWLEDGE

1. Draw a wave and label its: (Level 1–2)
   a. wave length
   b. amplitude
   c. trough.
2. State how: (Level 3–4)
   a. sound reaches your ear from someone clapping nearby
   b. light reaches your eye from a light bulb.
3. Outline how the ear detects sound. (Level 5–6)
4. a. Describe the differences and similarities between digital and analogue signals and explain why digital signals provide better quality communication.
   b. Describe why the use of optical fibres for bringing internet and TV into houses is increasing. (Level 7–8)

APPLYING SCIENTIFIC KNOWLEDGE AND UNDERSTANDING TO SOLVE A PROBLEM

5. Calculate the wavelength of a wave if four complete waves are 20 metres long. (Level 1–2)
6. Calculate the frequency of a wave if four complete waves are produced in 10 seconds. (Level 3–4)
7. Describe two ways soldiers could communicate with each other before the invention of electricity. (Level 5–6)
8. Two students are hiking near a cliff and get lost. They shout 'Help!'. However, all they hear is their own echo 2.3 seconds later. How far away are the students from the cliff? Take the speed of sound as 340 m/s. (Level 7–8)

INTERPRETING INFORMATION

9. Compare the properties of the four different sounds shown in the oscilloscope in Figure 6.31. (Level 1–8)

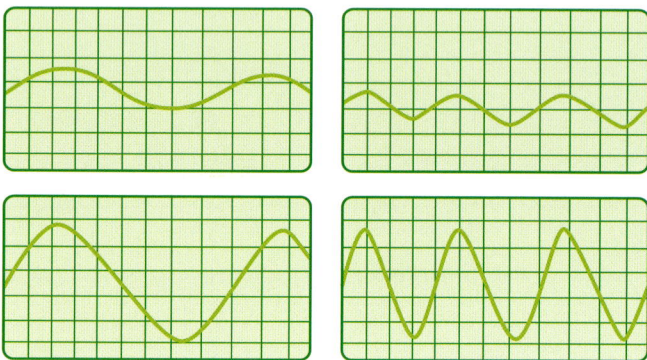

FIGURE 6.31 Four different sounds

10. A friend has started buying old vinyl records. She claims the sound is much better than from CDs. What would be your response? (Level 1–8)

REFLECTION

1. In its definition of communication, the IB includes the following statement.

   Communication involves the activity of conveying information or meaning. Effective communication requires a common 'language' (which may be written, spoken or non-verbal).

   a. How does your use of modern communication technology make communication easier or more challenging?
   b. Why is effective communication important to scientists?
2. Describe how waves are a method of transfer of energy.
3. In what other ways can energy be transferred?
4. a. Discuss how important you feel the development of modern communication technology has been to society over the last 20 years.
   b. What consequences of its use concern you?
5. Is digital always better than analogue?

UNIT 7

ASSISTIVE TECHNOLOGY

KEY CONCEPT
Systems

RELATED CONCEPTS
Energy

Form

Function

Equity

GLOBAL CONTEXT
Fairness and development – an exploration into how technology can provide equal opportunities for people with disability

STATEMENT OF INQUIRY
With appropriate design (and attention to form and function), technology can be used to ensure equal opportunities for people with disability.

INQUIRY QUESTIONS

FACTUAL
1. How do levers work?
2. What are the categories for classifying disability?
3. What are the names of five different types of simple machines?

CONCEPTUAL
4. What is the scientific principle behind why all these machines are useful to us?
5. How are work and force related?
6. How does the gearing in bicycles work?

DEBATABLE
7. To what extent do we need to include people with disability in all aspects of life? Is it possible?
8. Is a 100% efficient machine possible?

Introduction

Our interactions with the physical world require us to make things move or to hold them still. For example, we cut food with a knife, play on a seesaw and ride a bicycle. Increasingly, we use all sorts of systems that control machines so that we can carry out such activities as washing clothing, digging trenches and changing car tyres. Imagine what it would be like if you could not use your arms and legs, your natural body machines, to hold or move things. For some people, this is the reality of their daily lives. **Assistive technology** – machines and control systems designed for use within planned and unplanned spaces – makes the lives of people with a disability more productive and enjoyable, and attempts to create a society with greater equity for all.

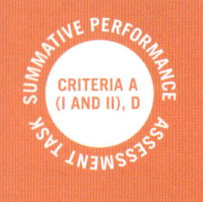

SUMMATIVE PERFORMANCE ASSESSMENT TASK
CRITERIA A (I AND II), D

Application of assistive technology

Prepare an illustrated report for a magazine about the challenges faced by people with a particular physical or movement disability. Outline how design and technology are used to enable someone with this disability to function more effectively in society. Focus on a specific problem and explain how science and technology is addressing it. You may choose to limit your report to a specific location such as your school, a library or a home.

Include discussion about the issues people with this disability still experience with gaining access to the necessary assistive technology or other necessary support. You could try to make contact with people with this disability to listen to their stories and opinions.

ATL

AFFECTIVE: EMOTIONAL INTELLIGENCE, THE ABILITY TO SHOW EMPATHY

In your report, try to show that you have tried to 'put yourself in the shoes' of the person with a disability, that you understand their perspective. This is called showing empathy.

Understanding disability

A person with a **disability** is unable to operate in the usual way that most people can function. There is quite a range of activity that is considered **normal**. People with a disability usually operate outside that range. For example, a person may not be able to run very fast, or throw a ball or get high marks in school, but that does not mean they have a disability. However, if they cannot control or coordinate their muscle movements, or if they find thinking difficult, they are likely to have a physical, movement or learning disability (Figure 7.1).

We all move between indoors and outdoors. Usually, we do not notice the **transition** from one to the other. But if you are in a wheelchair, getting up the steps from the street and into a building can be challenging. A design feature such as a ramp could make a significant difference (Figure 7.2).

FIGURE 7.1 Having a disability does not mean you can't have fun!

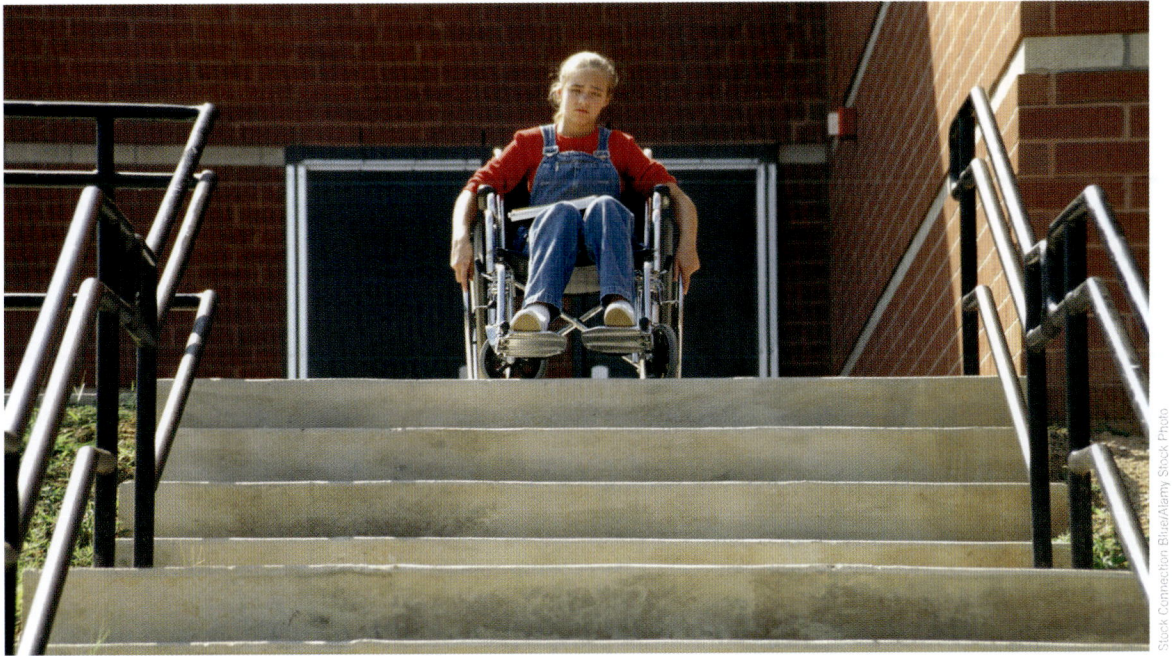

FIGURE 7.2 Assistive technology (wheelchair) and a pathway with an inaccessible design (stairs)

ACTIVITY: Disability classifications

To help understand and make decisions about disability around the world, the United Nations World Health Organization (WHO) has developed the **International Classification of Functioning (ICF)**. This classification recognises that people have many different aspects to their lives, and that their ability to share fully in all aspects may be affected by a disability. There are three dimensions that affect a person's full participation in life: structure/function, activity and participation.

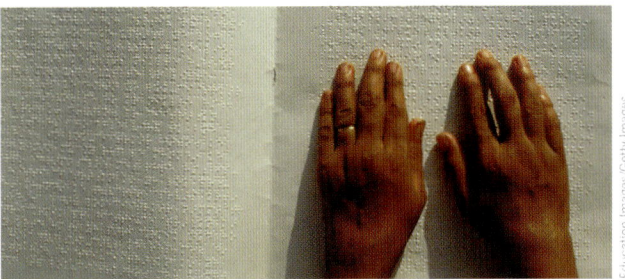

FIGURE 7.3 A blind person uses Braille to read books.

Disability can be grouped into categories. Find out about these categories on the weblink. Draw up a table as shown below and complete it for the eight categories of disability mentioned.

Category	What the category includes	Congenital/acquired	Effect on person

Go to http://mypsci3.nelsonnet.com.au and click on **Disability types** for more information on types of disability.

Structure and function restrictions

Disabilities arise from restrictions to a person's body **structure** and **function**. Vision impairments arise when a person's eyes (structure) cannot see (function). Similarly, people who cannot use their legs (structure) are unable to walk (function). What are the structure and function restrictions for a deaf person?

Activity-related restrictions

Some people cannot take part in everyday activities because of health issues, such as disease, chronic illness or abnormalities of body structure. Someone with chronic fatigue syndrome simply cannot get out and participate in normal activities. A person with a degenerative muscle disease, such as muscular dystrophy, may become exhausted from physical exercise much more quickly than their peers.

Participation restrictions

Some parts of society are closed or less accessible to people with certain disabilities. In some countries, people with an intellectual disability are excluded from casting their vote. Sometimes, a person is limited in their access to education because the school cannot provide sufficient supportive resources, such as ramps instead of stairs for people in wheelchairs.

Participation is affected by whether assistive technology is appropriate, accessible, available and affordable. A wheelchair may be available for a disabled person in a remote location, but it may struggle on rough terrain and become useless when the moving parts clog up with sand or mud. In a city, hoists may be available and easy to access, but too expensive to purchase or hire. There are many other ways that people with disabilities are restricted in participating in society.

FIGURE 7.4 Stephen Hawking has amyotrophic lateral sclerosis – a type of motor neurone disease, which confines him to a wheelchair. Despite this disability, he has made significant contributions to science.

(TA) PROVIDING OPPORTUNITIES FOR PEOPLE WITH DISABILITIES

Government rules and regulations are very important for people with disabilities. If there are no rules requiring schools, businesses and recreation facilities to cater for people with disabilities, then these people can be excluded from some activities. Find out what rules and regulations your school has to follow in order to provide an environment that makes it possible for people with disabilities to participate as fully as possible in school activities. If your school or your parents' workplace, or even places you visit often, such as your favourite museums, do not provide access for people with disabilities, you could take some action and petition them to change their rules and regulations in order to provide more access and opportunities for people with disabilities.

Go to http://mypsci3.nelsonnet.com.au and click on **Bionic eye implant** for more information on the bionic eye implant.

Recent developments in assistive technology

Research into blindness is so far advanced that blind people can now hope for an artificial implant into their eye that will improve their vision significantly. The ability to see is one of our most important senses. A person is legally blind if they can see less than 20% of what normal sighted people can see. In a number of countries, a bionic eye is being designed to improve vision for people with a condition in which the cells in the retina degenerate (stop working). An electronic chip will be surgically implanted onto the retina. That way it will stay fixed in place. Signals from a video camera mounted on a pair of spectacles will be transmitted to the chip to stimulate the cells in the retina. These signals can then travel along the optic nerve to the sight-processing centre in the brain. The first bionic eye implant took place in Manchester, UK in July 2015 with promising results.

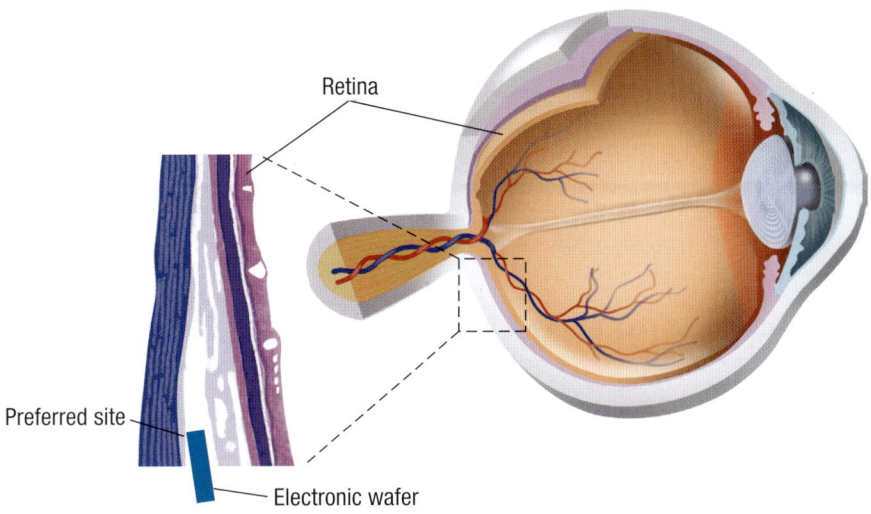

FIGURE 7.5 The bionic eye. Researchers hope to implant an electronic chip onto the retina so it can receive signals from an external camera and transmit them to the optic nerve.

A recent exciting development is the production of bionic hands with 3D printers. 3D printers are revolutionising the manufacturing industry. A bionic hand that may once have cost as much as 50 000 euros is now likely to cost more like 2000 euros.

Another development is the Quovis electric car (Figure 7.7), which allows people in wheelchairs to drive straight into the car and drive it. It is considered likely that the Google driverless car will also offer potential to support disabled people. A number of devices now help children with autism communicate via pictures. Other devices allow people with speech disabilities, such as those caused by paralysis, cerebral palsy and stroke, to communicate via artificial speech generated by recognition of eye movements. It has been suggested that Google Glass may also offer similar benefits to people with disability.

Go to http://mypsci3.nelsonnet.com.au and click on **Bionic hand project** to learn more about the bionic hand project.

Go to http://mypsci3.nelsonnet.com.au and click on **3D printing** to learn more about 3D printing.

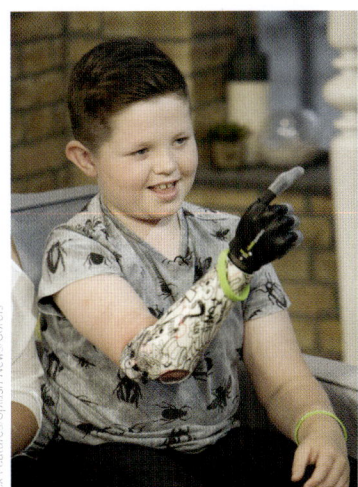

FIGURE 7.6 Nine-year-old Josh Cathcart, from Scotland, is the youngest person to be fitted with a special child-sized bionic hand. Josh was born without part of his right arm but now he can play with Lego® and even stick his thumb up thanks to the new state-of-the-art limb.

FIGURE 7.7 Quovis electric car from Vexel Automocion

Go to http://mypsci3.nelsonnet.com.au and click on **Assistive technology** to read more about different types of assistive technology.

Physical and movement disability

Some people are born with a physical or movement disability (**congenital disability**) and others become disabled during their life (**acquired disability**). When a person has a disability, whether congenital or acquired, they should still be able to enjoy life and contribute as fully as possible in family, education, work, relationships and recreation. Later in this unit, you will look at the role of technology in helping people with a physical or movement disability to enjoy and contribute to life.

Disabilities from wars

FIGURE 7.8 An Angolan landmine victim

Wars have disastrous effects on people and the environment long after peace has been restored. Land is made unproductive and dangerous. Imagine living in an environment where there are thousands of landmines and other explosive remnants of war. Imagine acquiring a disability by having your legs blown off by stepping on a landmine, years after a war has finished. This is the environmental reality for hundreds of thousands of people in Asia, Africa, Europe and South America (Figure 7.8). For instance, in Angola alone there remain 10 million land mines, and the country has an amputee population of 70 000. The Landmine Ban Treaty of 1999 has now been signed by 166 countries. It is estimated that the number of people dying from landmines has dropped from 20 000 to 4000 per year, which is still a large number. The WHO estimates that only 5% of people with disability receive rehabilitation in developing countries.

Go to http://mypsci3.nelsonnet.com.au and click on **ICBL** to visit the ICBL website.

TA INTERNATIONAL CAMPAIGN TO BAN LANDMINES

Click on the weblink to International Campaign to Ban Landmines (ICBL) to learn more about the tragic results of landmines used in wars. You and your class may decide to support ongoing efforts to stop the use of and to locate and remove landmines.

Go to http://mypsci3.nelsonnet.com.au and click on **Landmines video** to watch a video produced by the International Committees of the Red Cross.

REVIEW

1. State what is meant by the word 'disability'.
2. State an example of a physical or movement disability. How does this disability affect the person's life?
3. Outline an example of an assistive technology and explain how its design can impact negatively on its use.
4. Outline the three dimensions that can affect a disabled person's full participation in life.
5. State the structure and the function that are impaired for a deaf person.
6. Discuss two recent new examples of assistive technology and explain why they are potentially useful.
7. Discuss why landmines are one of the worst aspects of war.

Help from simple machines

A **machine** is anything that makes it easier to get something done. This includes all tools, from a pair of scissors to a complex computer. It even includes ramps. If a device helps us to get something done, it is called a machine. Most machines that we are familiar with are combinations of **simple machines**. We shall look first at simple machines and then see how they are combined into more complex machines.

The five simple machines that we will consider are:
1. inclined planes
2. levers
3. pulleys
4. wheels and axles
5. gear wheels.

Inclined planes

A sloping ramp, or **inclined plane**, allows you to raise or lower objects more easily. Pulling an object up a ramp is much easier than lifting it straight up. Also, you can more easily control the way something goes downhill using a ramp. The object is moved a greater distance, but less force is required to move it.

The energy you need to use to move something up a ramp is the same as the energy you use lifting it directly upwards. This **energy transfer** is called the **work** you do. Both work and energy transfer are measured in joules (J). Work and energy transfer are the product of the force (measured in newtons, N) you apply and the distance (measured in metres, m) over which you apply that force.

Work done = energy transfer
Work done = force applied × distance
Energy transfer (J) = force (N) × distance (m)

Ramps are used as an alternative to stairs to make it easier to move between levels. People in wheelchairs find it much easier to use a sloping surface than stairs to go up and down, both in and around buildings. You may have noticed the way ramps are used in your school to accommodate the needs of students in wheelchairs.

Wedges, chisels and screws are also examples of inclined planes. A screw is an inclined plane that spirals around to form a thread. Roads on mountains are made like really big screws – lots of ramps and turns as they spiral up to the top.

FIGURE 7.9 A mountain road is a series of ramps and turns, like a screw.

Using a ramp

A force of 100 N is applied to a 10 kg object to lift it directly upwards by 2 metres (Figure 7.10a). This requires 100 N × 2.0 m, or 200 J of work to be done.

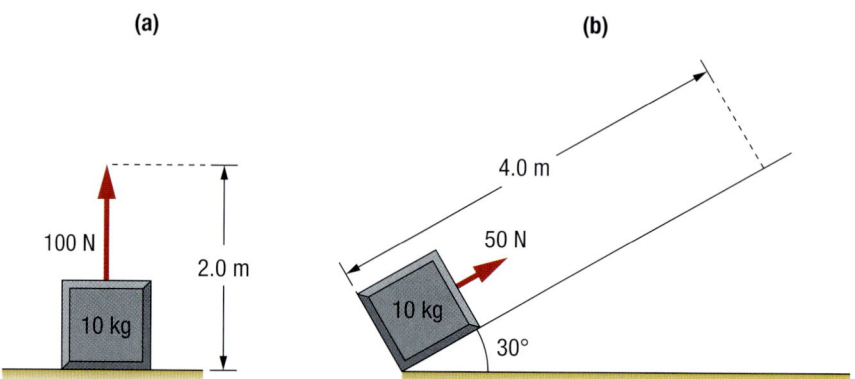

FIGURE 7.10 (a) Lifting an object vertically and (b) pulling the object along an inclined plane (ramp) to the same height

The same amount of work would be done if you pulled the object along a ramp by applying a 50 N force to it (Figure 7.10b). However, you would need a 4 m long ramp to do it: force (50 N) × distance (4 m) = 200 J of energy transfer.

In this example, we have assumed that all of the energy is transferred by the forces involved as they act on the object over a distance. There is another force – friction – that also acts on the object. Friction also acts over the distance and does work that reduces the amount of energy that would be transferred.

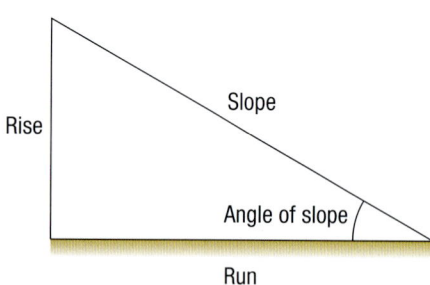

FIGURE 7.11 The gradient is the ratio of the rise (vertical height) to the run (horizontal distance).

Measuring the slope

The **slope**, or gradient, of a ramp is a measure of the angle the ramp makes with the horizontal ground. The greater the angle, the greater the slope. We measure the vertical height the ramp rises and the length along the ground it covers, and then divide them to get the ratio (see Figure 7.11).

Builders usually refer to slopes by a single number. A slope of 16 means that the distance is 16 units horizontally for every 1 unit vertically. You can see that a slope of 20 is less than a slope of 10. This is because a rise of 1 to a run of 20 (1:20) is less steep than a rise of 1 to a run of 10 (1:10). Draw a sketch to show this. Slopes for wheelchairs should be about 14 (1:14). For long slopes, horizontal sections are included to enable people to rest on the way up or to slow them as they come down.

Levers

A seesaw is a common lever. A lever is a rigid beam that rotates around something that does not move. The object or point around which the lever turns is called the **pivot** or **fulcrum** (Figure 7.12).

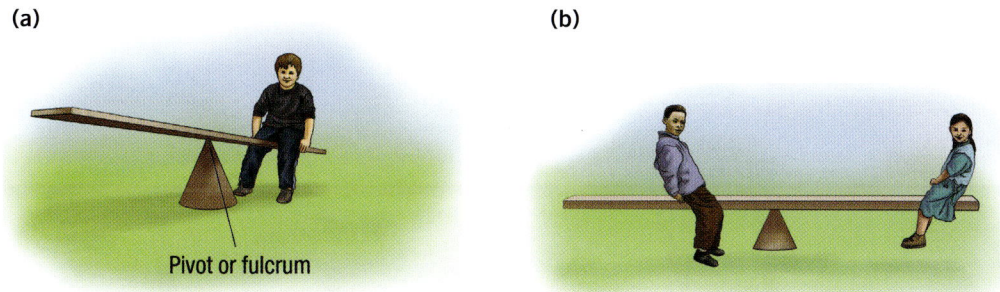

FIGURE 7.12 (a) One person on a seesaw makes it rotate. (b) Two people seated at the right spots can balance a seesaw.

On their own, each person on a seesaw causes it to rotate. But together, two people can make it stay still. In order to balance the seesaw, the heavier person has to sit closer to the pivot than the lighter person.

Levers are used to increase our effectiveness when moving an object. A heavy load on a lever can be moved as long as we can exert a force further away from the pivot. The force we exert is called the **effort**. The object we want to move exerts a **load** force. In Figure 7.13 you can see that only a small downwards force is needed on the beam to lift a very heavy object. The lever rotates but the fulcrum does not move.

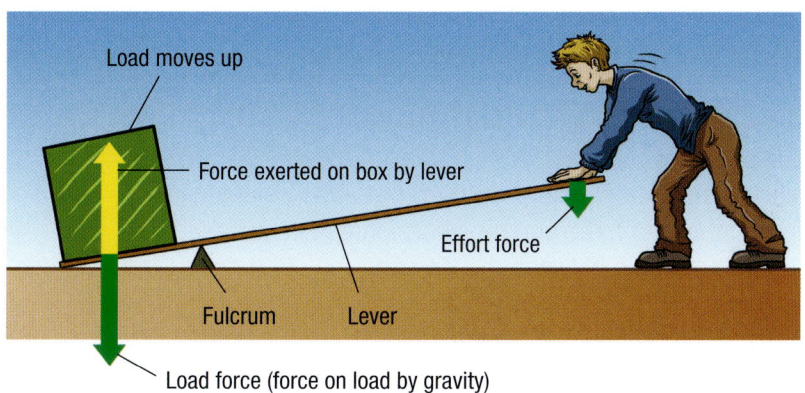

FIGURE 7.13 A lever at work

Moment

The turning effect of a force is called the **moment**. The moment depends upon both the size of the force and the distance between the force and the pivot.

Moment is calculated as follows.

Moment of the force = force × distance of force from pivot

Units: newton metre (N m) = newton (N) × metre (m)

Calculating moment

Here is an example of moment calculations.

Calculate the moment when a child of weight 500 N sits 1 metre from the pivot of a seesaw.

Moment = force × distance

Moment = 500 N × 1.0 m = 500 N m

ATL
COLLABORATION

Roles and responsibility when working in groups. How did your group approach the learning task? Were you able to accept the ideas others suggested? What roles did people play? Did you work out ways to resolve different answers between you? Did you help each other to learn effectively?

ACTIVITY

Looking at levers

WHAT TO DO

Working in small groups, look carefully at Figure 7.14.

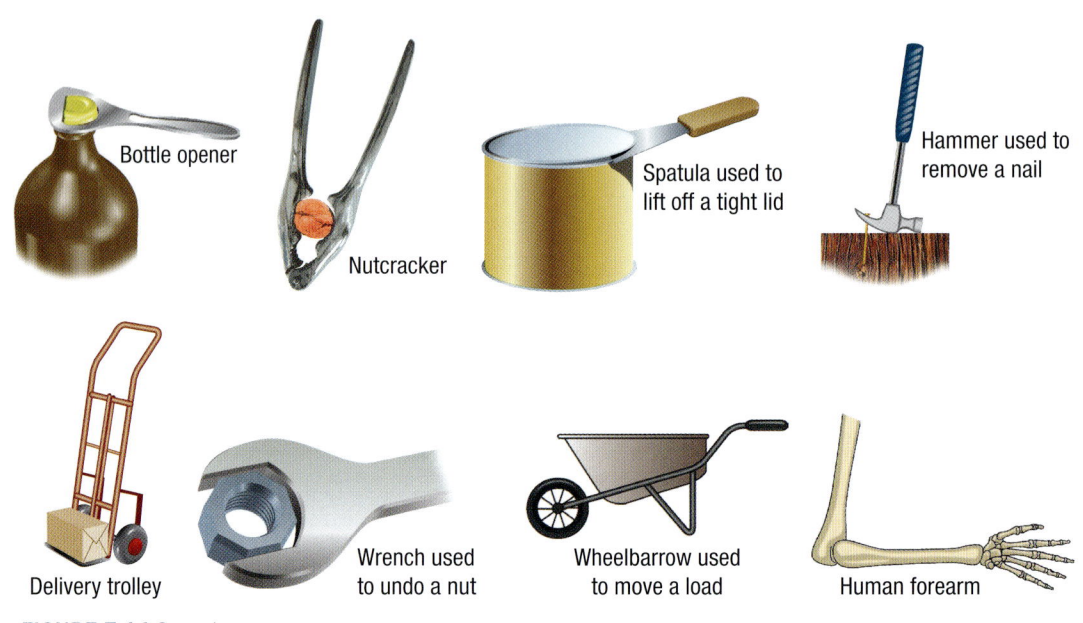

FIGURE 7.14 Some levers

In your group, answer these questions. What did you discover?
1 Which of these levers have you used?
2 Have you seen any of the others being used?
3 Identify the fulcrum, the load force and the effort force in each case.
4 In the seesaw the fulcrum was between the load and the effort. Is this true for each of these situations?
5 Does everyone in your group agree on where the effort, fulcrum and load are placed in each case?
6 Compare your findings with those of other groups.

When two children sit in a balanced way on a seesaw, the clockwise moment will equal the anticlockwise moment. This is called the **principle of moments**.

For instance, where would a child of weight 250 N sit on the seesaw to balance a child of 500 N who is sitting 1 metre from the pivot?

By common sense, you would probably say something like 'the child of 250 N weighs half the amount of the child of 500 N and so will need to sit twice as far from the pivot, that is, 2 metres from the pivot'. When the seesaw is balanced, the moments add up to zero.

Clockwise moment = anticlockwise moment

$$500 \text{ N} \times 1.0 \text{ m} = 250 \text{ N} \times \text{distance}$$

$$\text{Distance} = \frac{500}{250} = 2.0 \text{ metres}$$

Robotic arms

Your forearm is a lever. You should be able to locate the effort force, fulcrum and load force on your forearm when you use it to lift a book.

Artificial arms and legs have been used for some time. Artificial limbs contain levers too. Robotic arms can respond to the wearer's thoughts. For example, a paralysed person might want the arm to reach for food, so they just have to think about reaching for that food.

Robotic limbs work by having a microchip implanted inside the person's brain. The chip converts a thought into a signal. This signal is sent wirelessly to a computer in a backpack worn by the person. The computer turns the signal into a motor command. This command is sent wirelessly to a chip implanted in the person's arm. This second chip stimulates the nerves needed to move the muscles of the arm (Figure 7.15).

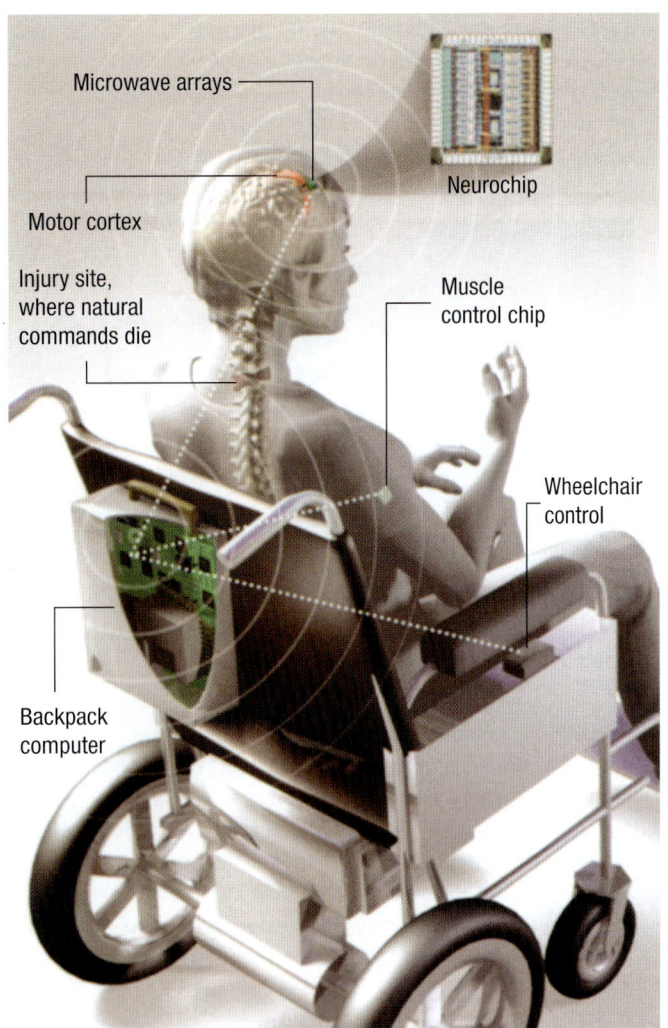

FIGURE 7.15 Human thoughts and artificial limb movement are connected via the brain and a microchip.

Pulleys

Pulleys and ropes can be used to lift things. They are common in factories, building sites and farms. A **pulley** is a wheel with a groove around the edge. The rope fits inside this groove. A pulley hung up high with a rope over it can be used to lift heavy objects. You pull down on one end of the rope to pull up the object on the other end of the rope (Figure 7.16).

Many pulleys can be used together to make objects even easier to lift. By looping the rope around several pulleys, you magnify your force. However, the object will only be lifted a small distance on each pull. This is similar to our discussion of inclined planes. The work done to transfer energy is the product of the force applied and the distance moved.

Look at the two-pulley system shown in Figure 7.16b. You can see that if the load is raised 1 metre, the effort will need to move 2 metres because of the arrangement of the ropes.

Go to http://mypsci3.nelsonnet.com.au and click on **Learning disabilities support** for tips on helping/supporting people with learning disabilities. If you have a friend who has difficulties with learning and following, these links and resources might help your other friends learn more about supporting them.

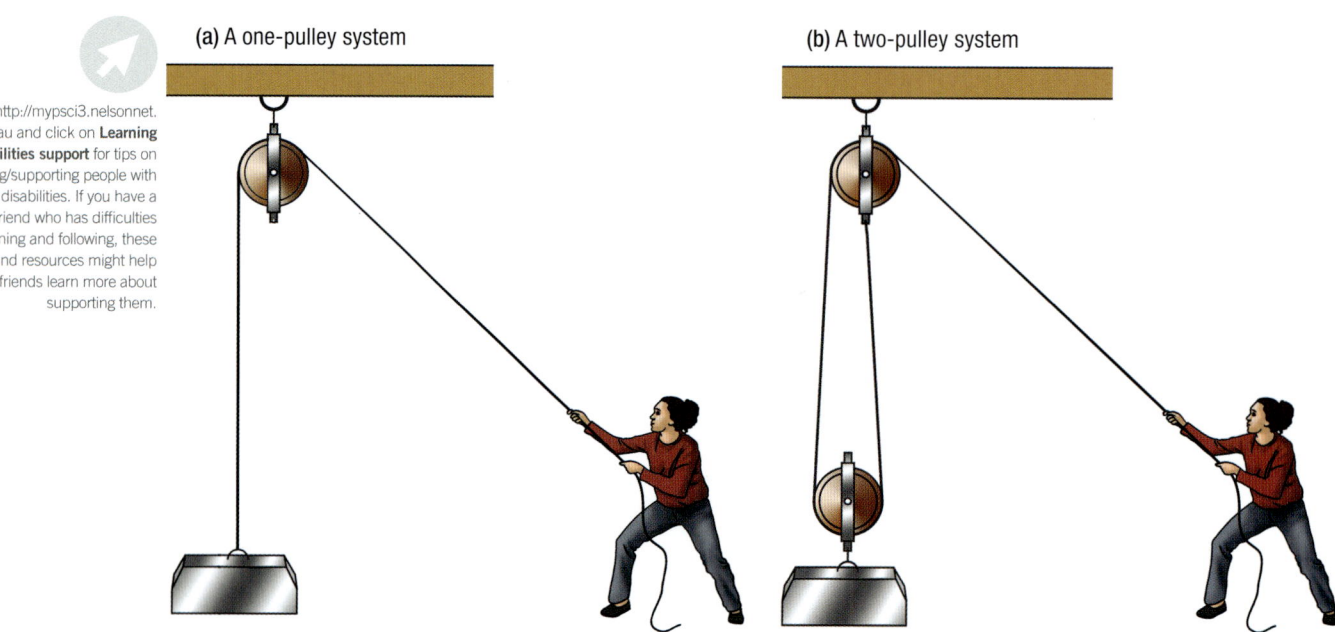

FIGURE 7.16 Pulley systems are used to lift a load.

Pulley calculations

What effort will be needed to lift a load of 100 N a distance of 1 metre using this two-pulley system?

Calculation:

The work carried out on the load will be 100 N × 1.0 m = 100 N m.

The work done by the effort will be the same (if 100% efficient):

$$100 \text{ N m} = \text{effort} \times 2.0 \text{ m}$$
$$\text{Effort} = \frac{100}{2} = 50 \text{ N}$$

With a two-pulley system, the ratio of the load to the effort is 2. What do you think will be the ratio of the load to the effort for a one-pulley system and a three-pulley system?

Wheels and axles

Probably one of the greatest inventions of all time is the wheel. Imagine life without wheels. An axle is the solid rod on which a wheel is placed. The wheel turns when the axle turns.

Many things use a **wheel and axle**, even door handles and screwdrivers. If you turn the steering wheel of a car around a large distance, its axle, which is inside the steering column, only turns around a little way.

There are some similarities between the seesaw lever and the wheel and axle. For a lever, the forces cause rotations around a point, the pivot. For an axle, the line down the centre is the axis (similar to the pivot); your hands rotating the steering wheel are the effort, a long way from the rotation axis; and the outside of the axle, the part you want to move, is the load. The load moves a small distance while the steering wheel rotates a bigger distance (Figure 7.17).

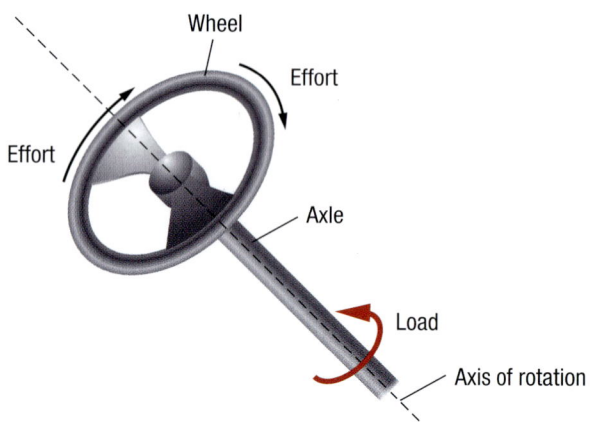

FIGURE 7.17 Steering wheel and axle: a small effort on the steering wheel a long way from the axis causes the axle to turn to a lesser extent, but sufficiently to steer a vehicle. The load is shown in opposition to the rotation required.

Gear wheels

A **gear wheel** is a wheel with teeth around it. The teeth of one gear wheel fit into the teeth of a second gear wheel. When the first gear wheel turns, so does the second. In this way, gears transfer movement. A **driving gear** is the gear wheel that is turned around by a motor or by someone pushing it. A **driven gear** is a gear wheel that is turned around as a result of the driving gear turning (Figure 7.18).

In Figure 7.18, the two gear wheels have the same number of teeth. This means that when gear A turns around once, gear B will turn around once too. However, it will turn in the opposite direction.

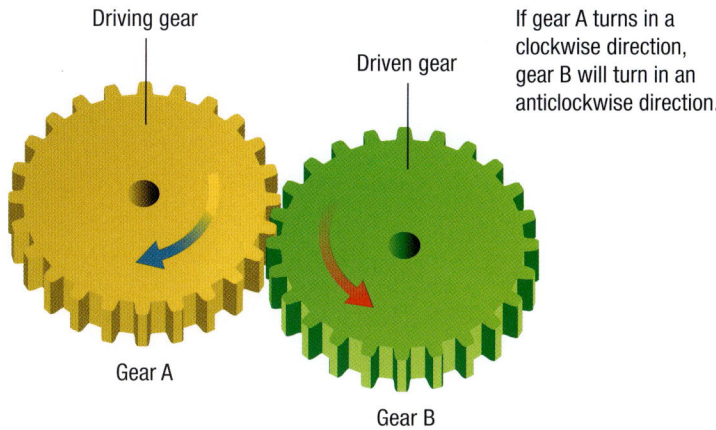

FIGURE 7.18 Toothed gear wheels. The teeth fit together to transfer movement from one wheel to the other.

Gear wheels can also be used to change the speed of an object. In Figure 7.19, you can see that the driving gear (C) has three times the number of teeth of the driven gear (D). A fan is attached to the axle of gear D. By the time gear C has rotated once, gear D will have completed three rotations. This means that if the motor turns gear C around 50 times per minute, the fan will turn around three times faster at 150 times per minute.

The relationship between the number of teeth on two connected gears is called the **gear ratio**. In many applications, such as inside wind-up watches and clocks, many gear wheels can be linked together, each going at a different speed.

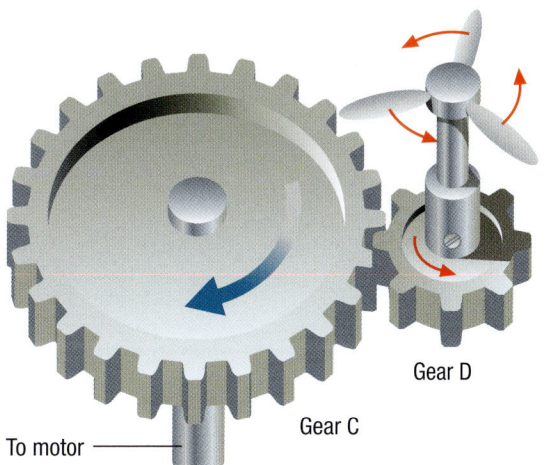

FIGURE 7.19 Changing the speed of rotation with gear wheels

Sometimes gear wheels are not even touching. Instead, a chain links them. This allows the two gear wheels to be a distance apart. A bicycle is a typical example of the use of gear wheels connected by a chain. The pedal is connected to one gear wheel. The chain connects this gear wheel to another on the back wheel (Figure 7.20). If you are a serious rider, you might have a bike with many gear wheels, so you can choose a gear best suited to the surface you are riding on.

There are many examples where levers, pulleys, wheels and axles, and gears are used in assistive technologies.

FIGURE 7.20 A chain connects the gear wheels on a bicycle.

CRITICAL THINKING
Consider the use of graphs to show results, different kinds of graphs, choice of axes, drawing of lines of best fit, and possible use of uncertainty bars.

The functioning of bicycle gears

INVESTIGATION 7.1

YOUR CHALLENGE
Design an investigation into how the gear ratio chosen for a bicycle affects the movement of the wheels.

THIS MIGHT HELP
You will need a bicycle with gears on both wheels. Your experiments should use at least two different gears on the front wheel, and a variety of gears on the back wheel.

Carry out and write up your investigation following the guide in Appendix 3 on page 177 or as advised by your teacher.

REVIEW

1. Name five simple machines.
2. Draw a diagram to illustrate the difference between a slope of 20 (1:20) and a slope of 10 (1:10).
3. On a diagram of a human forearm show the position of the pivot, the load and the effort. What kind of simple machine is this?
4. A car mechanic uses a wrench to loosen a nut before changing a car tyre. He applies a force of 100 N at a distance of 0.5 m from the wheel nut, which acts as the pivot. Calculate the moment about the pivot.

5. Two children sit on a seesaw to make it balance. The older child of 300 N weight sits 0.5 m from the pivot. Where will her younger brother who weighs 150 N have to sit to balance the seesaw?
6. What machine could be used to lift something up by pulling downwards? Provide an illustration.
7. A machine made up of two pulleys is used to lift up a load of 200 N through a height of 2.0 m.
   a. How much work is done in lifting this load?
   b. How far will the effort move during this lifting?
   c. What will be the value of the effort needed?
   d. Explain the principle behind how this two-pulley machine works.
8. Illustrate the use of a machine that changes the direction of rotation of a wheel.
9. A 200 N force is applied to a bucket of water to lift it 2.5 m vertically up from a well and onto the ground. The bucket is then hauled 7.5 m along a slope until it is 2.5 m above the ground.
   a. Sketch this scenario.
   b. How much work was done to raise the bucket out of the well and onto the ground?
   c. How much energy was needed to move the bucket up the slope?
   d. What force was applied to the bucket along the slope?

Electrical control systems

Levers, pulleys, wheels and axles, and gear wheels can all be operated by electric motors. In these cases, only the flick of a switch is needed to get things moving. A person with limited mobility can be lifted and moved in a sling attached to a hoist. The hoist is a set of linked levers. The effort is supplied by a rod that is driven by an electric motor.

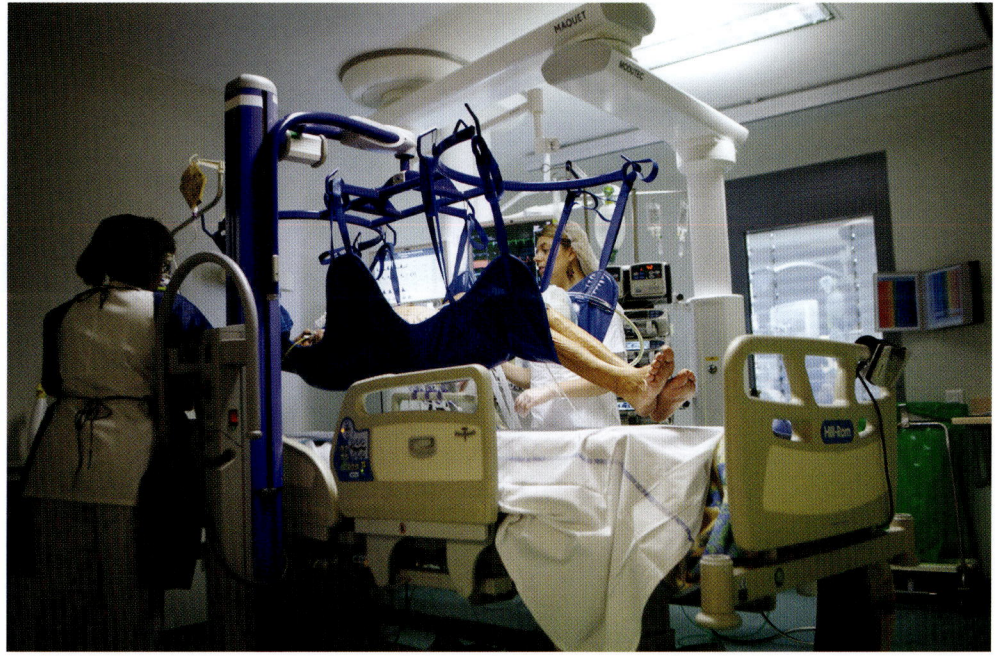

FIGURE 7.21 A mobile hoist made from levers is electronically controlled.

Pneumatic switch

People with severe disabilities can do amazing things with assistive technology. A person with a severe disability (for example, someone whose strongest function is breathing) can use their breath to turn a switch on or off. This **pneumatic switch** uses the force applied by an outward breath to turn a switch on and an inward breath to turn it off. In this way, they can control the movement of equipment, such as a wheelchair, sailboat or computer mouse (Figure 7.22). On a boat, a sailor can use a pneumatic switch to steer with the tiller, and to raise and lower the sails and position them to use the wind effectively.

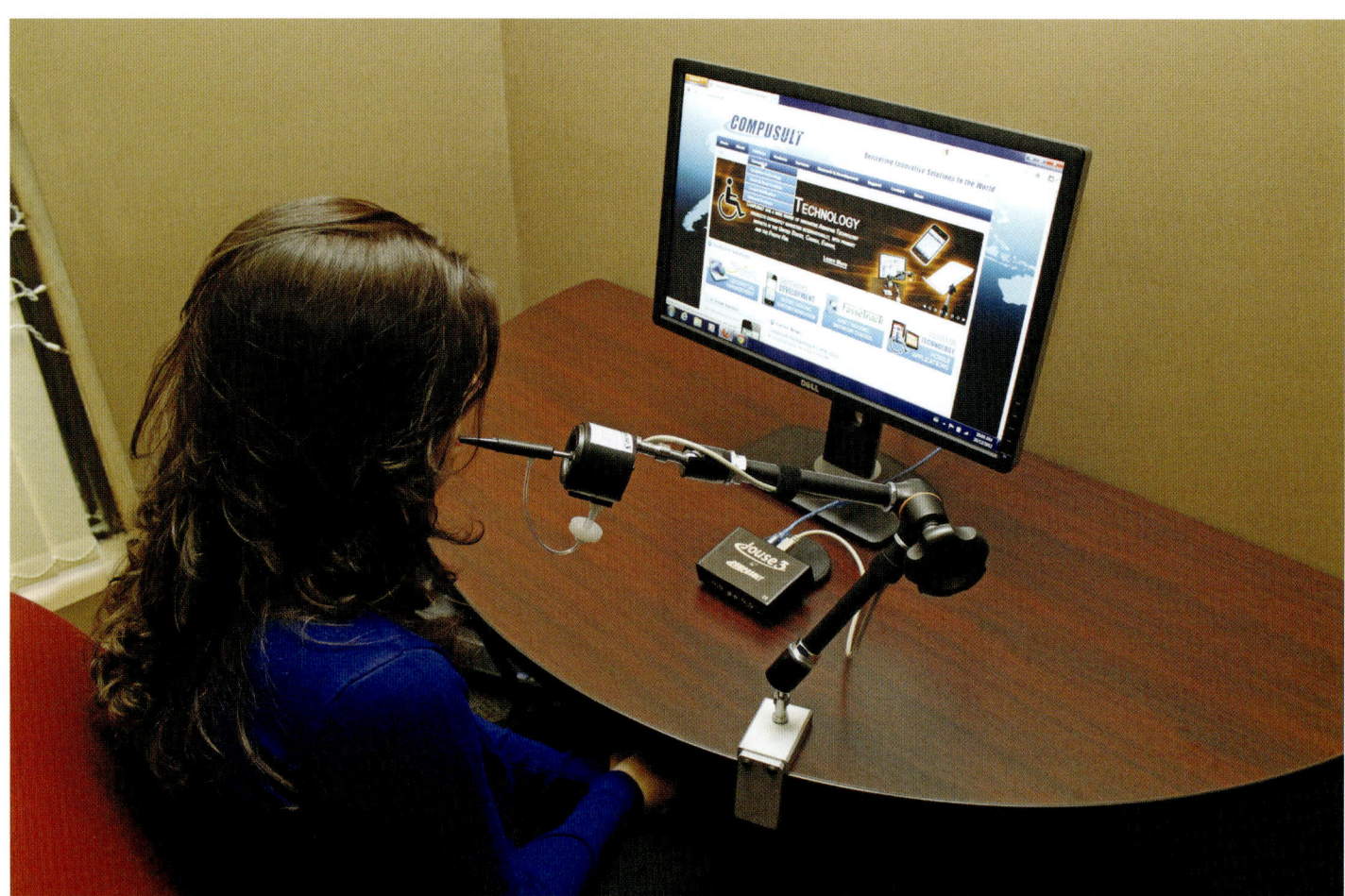

FIGURE 7.22 A joystick mouse operated by pneumatic switches

| ACTIVITY | **Technology equals freedom for high-quad sailor** |

Terry LeBlanc was a land surveyor who became a **high quadriplegic** in a diving accident – he now has no muscle control below the neck. Nevertheless, Terry is able to go out on the water and sail. He even competes in races. Part of the joy for Terry is working the entire experience by intuition as much as by mariner's skills. He merely needs to bite, breathe and think to sail his boat and gain an intense feeling of freedom on the waters of English Bay in Vancouver, Canada.

FIGURE 7.23 Terry LeBlanc in his sailing boat

Terry uses 'sip-and-puff' technology on his Martin 16 sailboat. He sips on the ergonomically shaped mouthpiece to head the boat to starboard when he's in the tiller mode and he puffs (or exhales) when he wishes to turn to port. He bites on the straw when he wants to go into 'sail' mode to trim, tack or sail against the wind. The sip-and-puff system is a dual-switch system that uses pneumatic technology connected to clever electronics. A small sip or gentle puff of air is all that is required to activate this switch forward or back. A single piece of tubing, accessible to the user, controls both switches. All functions, including setting the sails and steering the boat, are controlled by the sailor's breath. The term 'sip-and-puff' is somewhat misleading. No actual sipping or puffing, to the degree normally thought of as sipping or puffing, is required. Unlike other control methods, sip-and-puff control keeps conductive material away from the user. This eliminates the possibility of electrical shock. 'It's quite simple,' Terry said recently. 'Now in a racing situation, when you have to jockey about for position at the start line, it can get tricky. But I find it really quite intuitive.' Terry has made many friends through his sailing, either through racing or through the social activity that is synonymous with boating. 'The whole social aspect is quite neat,' he said. 'I like to hang out on the second floor deck at the sailing centre, and you can never tell who you will meet.'

Sam Sullivan Disability Foundation

1. Use the article to create a glossary of four new words.
2. Terry is a high quadriplegic.
   a. How did this happen?
   b. What effect did this have on his body?
3. Give four reasons why Terry enjoys sailing.
4. State the meaning of 'tiller mode'.
5. State what a pneumatic switch is. How is it made to work?
6. a. How do we normally refer to 'sipping and puffing'?
   b. Explain why 'sip-and-puff' is not an accurate description of the controls on Terry's boat.
7. Does Terry's breath directly act on the tiller or sails? Explain.

Global navigation satellite systems (GNSS)

Go to http://mypsci3.nelsonnet.com.au and click on **GPS devices** to see recommendations for GPS devices for people with Alzheimer's.

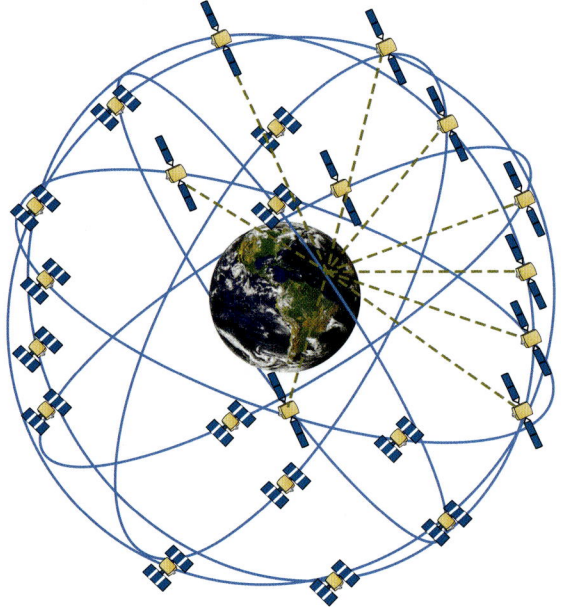

FIGURE 7.24 Your signal from Earth enables medium-orbit satellites to pinpoint your position.

We use **latitude** and **longitude** to describe our position anywhere on Earth. By sending a small, electronic signal to a system of satellites, our position can be fixed to within a few metres. The satellites are in orbit about 20 000 kilometres above the Earth. In order to cover the whole globe, 20–30 satellites are needed. When you send a signal, it arrives at the nearest satellites (some satellites are behind the Earth, so out of range, from our point of view), which then sort out the information and send it back as your position. The more satellites that process your signal, the greater the accuracy of your position. The US GNSS is called the Global Positioning System (GPS); the Russian GNSS is called GLONASS. The European system, Galileo, will be ready around 2020.

A person with a disability may need help quickly if they are lost or get into trouble. A special-purpose mobile phone or a similar device kept somewhere on their body (for instance in a shoe) with a global navigation satellite system (GNSS) tracker can be used to locate people quickly, and allows carers to send supportive messages.

New devices based on GPS are also available for blind or visually impaired people to give them continuous feedback about their position or to provide directions. Recently, a group of blind and visually impaired people walked through the Vosges mountains in France with nothing but a GPS device and walking sticks.

Go to http://mypsci3.nelsonnet.com.au and click on **Blind hikers** to see a video showing these blind walkers in the Vosges mountains.

FIGURE 7.25 This blind boy is using a GPS device for guidance and a white walking stick.

ACTIVITY — Design a machine for a person with a disability

Think about the different types of disabilities that people might have. Select one of the challenges they might face from the time they get up in the morning to the time they go to bed at night. How could you help them?

Entrepreneurs see opportunities that other people do not see. If you design or work out ways or opportunities to include other people, and what you have is unique and innovative, you might have the opportunity to either sell your products, or sell your patent rights, thus making an important contribution to the lives of many people.

Use the work on machines covered in this unit, or other ideas discussed in the unit, to design your ideal machine that will help a person with a disability to perform a particular action or that will help them all day to do most of the things that people without disabilities can do.

Let your imagination loose!
Draw the machine.
Describe what it can do.

ATL

CREATIVITY
Enjoy being creative and imaginative, designing new products, and showing entrepreneurial skills.

REVIEW

1. Outline two situations where pneumatic switches could be useful for disabled people.
2. Describe the orbits of the satellites in a GNSS system.
3. Describe a possible use for a GPS device for someone with Alzheimer's disease.
4. Outline what the term 'entrepreneur' means.

UNIT QUESTIONS

CRITERION A

EXPLAINING SCIENTIFIC KNOWLEDGE

1. State what is meant by the moment of a force. (Level 1–2)
2. State an example of how an assistive technology is related to a particular disability. (Level 3–4)
3. How does a steering wheel make use of a simple machine? Draw a diagram that shows the axis of rotation as well as the place where the effort and load are located. (Level 3–4)
4. Name and outline each of the three dimensions used by the United Nations to help understand disability. (Level 5–6)
5. Outline how a GNSS works. (Level 5–6)
6. Describe in some detail a scenario in which a system using GNSS technology would be of assistance to a person with a disability. Include a discussion of any problems you imagine could be experienced by its use. (Level 7–8)
7. Describe with examples the advantages of using pulleys and gears in machines. (Level 7–8)

APPLYING SCIENTIFIC KNOWLEDGE AND UNDERSTANDING TO SOLVE A PROBLEM

8. What piece of equipment could you use to try to remove the lid from a tin of paint? (Level 1–2)
9. Why is it easier to shut a door by pressing on the handle near the edge of the door than by pushing near the hinge of the door? (Level 1–2)
10. Calculate where a child of weight 200 N should sit on a seesaw to balance a child of weight 300 N who is sitting 2.0 m from the pivot. (Level 3–4)
11. Outline how sip-and-puff technology could be used for a school student in a wheelchair. (Level 5–6)
12. Compare the force needed to push an object of weight 60 N up a 2.0 m long slope to a height of 1.0 m with the force needed to lift the same object vertically by 1.0 m. (Assume no friction is involved, 100% efficiency.) Use diagrams to illustrate your answer and show your calculations. (Level 5–6)
13. What would be the result of choosing a gear with 50 cogs on the front of your bike, with a gear with 10 cogs on the back? Describe what you would do if this required too much effort going up a hill. (Level 7–8)
14. Describe why using a two-pulley system makes it easier to lift an object than a one-pulley system. Explain this in terms of energy. (Level 7–8)

INTERPRETING INFORMATION

15. Here are some results from an experiment that used different numbers of pulleys to lift an object. (Level 1–8)

Number of pulleys	Force required (N)
0	6.0
1	6.1
2	3.5
3	2.8
4	2.2

a. Draw a line graph to show these results with a line of best fit.
b. What conclusion could you make from these results?
c. Is the relationship directly proportional?
d. Are the results as you would expect? Explain.
e. What errors can you imagine in this experiment?
f. How could you reduce these errors?

REFLECTION

1. In this unit, we considered how machines transfer energy.
   a. Describe the relationship between energy and work.
   b. Describe the energy transformation that takes place when a person uses a pulley to lift an object.
   c. Explain how a smaller force can be used to lift an object with a bigger weight (force).
2. When designing a new device the relationship between form and function is important. How does this idea apply to the design of bicycle gears?
3. Do you feel in your country there is equity in the opportunity for people with disabilities to participate in society? Give some examples of where improvement is needed.
4. Discuss whether a machine can be 100% efficient.

UNIT 8

THE SUN AND MOON IN OUR LIVES

KEY CONCEPT
Systems

RELATED CONCEPTS
Models

Cycles

Interactions

GLOBAL CONTEXT
Orientation in time and space – how human communities are affected by the Sun and the Moon

STATEMENT OF INQUIRY
Human communities are deeply connected to the cycles in nature caused by the interaction of the Sun, Moon and Earth.

INQUIRY QUESTIONS

FACTUAL
1. What are the main reasons that Earth experiences seasons?
2. What happens during eclipses and tides?

CONCEPTUAL
3. How do the Sun, Earth and Moon interact?
4. Why are the seasons, tides and phases of the Moon all cyclic?
5. How do weather conditions influence the communities we live in?

DEBATABLE
6. Are we as people a product of the climate and the seasons?

Introduction

Since human beings first wandered the planet, they have felt the comforting warmth from the Sun in the day and looked into the sky at night with wonder to see the Moon and stars. The influence of these objects in the sky on the human mind can be seen in art, language, monuments, religion, music and science through the centuries. Human understanding of the objects in space has changed a lot over time; human ingenuity has let us find out much about objects that are so far away. Only in very recent times have we developed the technology to let us leave the planet, step on the Moon, closely examine the Sun, and send space probes deep into the solar system to keep on looking for answers to some of the questions we still have about life, the universe and everything. Life on Earth depends on the Sun, but we can equally argue that life on Earth also depends on the Moon. We can say that we don't just live on Earth, but that we live on a linked system of Earth and the Moon.

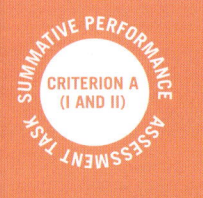

COMMUNICATION
Skilled public speaking

Human communities are deeply connected to the cycles in nature caused by the interaction of the Sun, Moon and Earth

Show your understanding of this Statement of Inquiry by writing an essay of 300–500 words on one of the following topics. Read your essay to the class, using your best public speaking skills.
- How does the presence of the Moon influence Earth?
- How has the influence of the Moon been perceived in various cultures and religions?
- How does the Sun affect communities on Earth?
- How do tides affect life on Earth, people, other organisms and communities?

Influence of the Sun on Earth

The Sun is the source of most of our energy, for plants to photosynthesise, for our food, for our weather. Without stars like our Sun, planets could not form; nor could they have **orbits**. The Sun is massive. It has a mass of 2 million million million million million kilograms (2×10^{30} kg). It is an enormous ball of gas and **plasma** undergoing **nuclear fusion** reactions that release huge amounts of energy into space. Each second, the Sun produces enough energy to meet all of the power needs of the USA for 9 million years.

The Sun is also spinning on its own axis. Four hundred years ago, Galileo Galilei observed a dark spot on the Sun called a **sunspot**. He noticed it disappear from one side of the Sun, and a few days later observed the same sunspot appear on the opposite side. He concluded that the Sun must be rotating.

Earth orbits the Sun once every 365 and a quarter days. Earth's orbit is not a perfect circle; it is a slightly squashed circle called an **ellipse**.

That means that at some times Earth is slightly closer to the Sun than at other times. For example, Earth is slightly closer to the Sun in January

FIGURE 8.1 Our Sun in its glory.

than it is in July. It is a common mistake for people to believe that this is the explanation for the different seasons.

The Sun in ancient mythology and religions

Sun is called Sol or Helios in some cultures. Not surprisingly, in most ancient cultures the Sun featured as a very important god, a solar deity associated with power and strength. Some people were Sun worshippers. Solar deities included Magec (Africa), Wala (Australia), Malakbel (Arabia), Nanauatzin (Aztec), Marici (Buddhism), Etain (Celtic), Ra (Egypt), Apollo (ancient Greece) and Helios.

Maui is a Polynesian solar deity. The story goes that the days were too short for his mother to make bark cloth, so he cut off tresses from the dress of his wife Hina to make a rope to catch the Sun as it was rising. He then beat the Sun with the magic jawbone of his grandmother, and so the Sun became so weak it could only crawl along its course across the sky.

A common image in many cultures was the Sun travelling across the sky in a chariot or boat. In Hinduism the Sun is seen as a visible form of god. The idea of Sol Invictus (unconquered Sun) comes from the ancient Greeks and became an important cult in Roman times. Before it was taken as the birth of Christ, 25 December, around the winter equinox, was celebrated as the birth of Sol Invictus.

Solstices (the shortest and longest days of the year) have been reasons for special festivals in many cultures, and also the time of mating of animals, sowing of crops and checking winter reserves of food. Neolithic archaeological sites such as Stonehenge in the UK and Newgrange in Ireland are examples of temples that appear to demonstrate an understanding of the solstices.

FIGURE 8.2 Chinese Sun goddess Xihe on a chariot with the Sun, pulled by a dragon

FIGURE 8.3 Greek and Roman Sun god Helios, at the temple of Athena at Troy

FIGURE 8.4 Stonehenge – built 2500 BCE. It is aligned to the morning sunlight from the summer and winter solstices.

Influence of solar storms on Earth

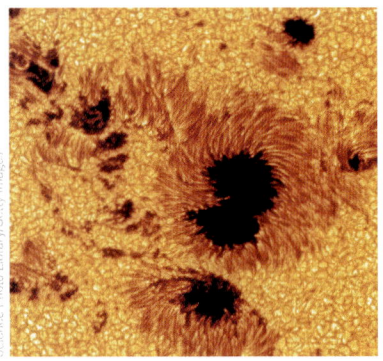

FIGURE 8.5 Sunspots, like the ones observed by Galileo, can be seen in this photo.

The first recorded person to observe sunspots on the surface of the Sun was the Chinese astronomer Gan De in 364 BCE. Sunspots appear as dark areas because they are cooler regions, and occur where there is intense magnetic activity. Sunspots occur in cycles of approximately 11 years as the magnetic field of the Sun changes.

NASA's Solar Dynamics Observatory enables scientists to study solar storms that disrupt communications on Earth. Solar particles interacting with Earth's magnetic field create the colourful Northern and Southern Lights. They can also create problems with **satellite** and radio communication. In 1989, solar storms caused electricity blackouts in Quebec, Canada, which affected millions of people.

Solar Dynamics Observatory

ACTIVITY

To learn more about the Solar Dynamics Observatory (SDO) that NASA launched in 2010, go to the Stanford Solar Center weblink. There is a wealth of interesting information on the sites about the SDO project and the Sun generally.

1. Choose one topic from the website that is of interest to you and write a 300-word summary of what you learnt.
2. Open the SDO Data in the Classroom PDF file.
    a. Watch the video mentioned on page 7 and complete the associated quest worksheet.
    b. Watch the videos mentioned on pages 8–16 and complete the activities that your teacher suggests.
3. On the Stanford Solar Center site, you will find the spinning Sun activity, in which you will be guided on how to use photographs of sunspots to measure the speed of rotation of the Sun. Did you know the Sun was rotating?

Go to http://mypsci2.nelsonnet.com.au and click on **Stanford Solar Center** to complete the activity on the SDO.

Measuring the diameter of the Sun using a pinhole camera

EXPERIMENT 8.1

PERFORMANCE ASSESSMENT TASK
CRITERION C (I, II, IV AND V)

MATERIALS
- sheet of card
- sheet of white paper or a screen
- blackout for a window (e.g. black card that can be taped over a window)
- sticky tape or masking tape
- aluminium foil about 3 cm × 3 cm
- pin or sharp point
- ruler
- scissors
- candle or other light source

PROCEDURE
1. Cut a square about 2 cm × 2 cm in the centre of the sheet of card.
2. Put the aluminum foil over the 2 cm square hole and tape it to the sheet of card.
3. Use a pin or similar to make a small hole in the foil.

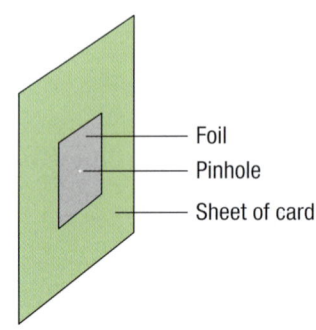

FIGURE 8.6 The pinhole camera

4. Your sheet of card complete with pinhole is your 'pinhole camera'.
5. Check that you get an image with your pinhole camera.
   a. Place a lit candle (or other light source) in front of the pinhole and then turn off all other lights.
   b. Hold a sheet of white paper or a screen a few centimetres from the back of the camera. You should be able to see an image of the candle flame (or other light source) on the screen.
6. Black out a window that faces the Sun, and cut a square about 2 cm × 2 cm in the centre of the blackout material.
7. Put the foil over the square hole and tape it to the blackout material.
8. Use a pin or similar to make a small hole in the foil.
9. Turn off all the lights in the room.
10. Hold a sheet of white paper or a screen as far from the pinhole as possible. You should now be able to see an image of the Sun on the paper or screen.
11. Measure accurately and record the distance between the pinhole and the white paper or screen.
12. Measure accurately and record the diameter of your images of the Sun.
13. Calculate the real diameter of the Sun from your measurements.

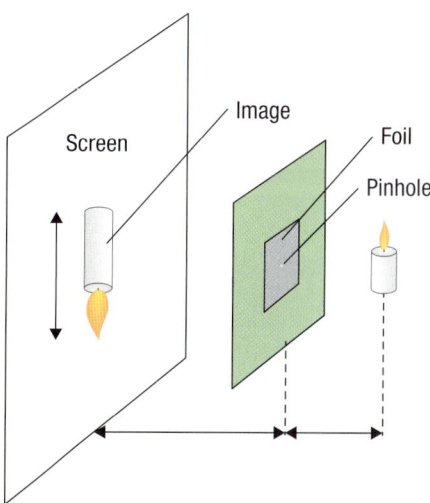

FIGURE 8.7 Finding the diameter of the Sun

RESULTS
The distance between my image and the pinhole camera was _____ cm.
The diameter of my Sun image was _____ cm.

CALCULATION
- The diameters of the Sun and the image are in the same proportion as the distances between the Earth and the Sun and between the screen and the pinhole.

$$\frac{\text{Diameter of Sun}}{\text{Diameter of image}} = \frac{\text{distance from Earth to Sun}}{\text{distance to screen from pinhole}}$$

$$\text{Diameter of Sun} = \frac{\text{distance from Earth to Sun} \times \text{diameter of image}}{\text{distance to screen from pinhole}}$$

- The distance from Earth to the Sun is 149 600 000 km
- Make sure that you change all your distance measurements to the same units, e.g. metres.
  Final answer =

EVALUATION
How close is your result to the accepted value? How much variation was there in the results of other groups?

Discuss the errors in the experiment, and suggest how the experiment could be improved.

SAFETY
Never look directly at the Sun. It can cause serious eye damage.

INVESTIGATION 8.1

The size of the image in a pinhole camera

YOUR CHALLENGE
You have just used a form of a pinhole camera in Experiment 8.1. This time your challenge is to investigate how one or more variables affect the size of the image from a candle (or another light source).

CRITICAL THINKING
The ability to formulate and explain a testable hypothesis

THIS MIGHT HELP
Think about the different ways you can make the image bigger. You should produce some elegant graphs of your results.

Carry out and write up the investigation following the guide in Appendix 3 on page 177 or as advised by your teacher.

Explaining the seasons

Figure 8.8 shows how the combination of Earth rotating around the Sun and Earth spinning on its axis causes the seasons. During summer in the northern hemisphere, the tilt of Earth means the northern hemisphere receives the Sun's rays more directly. This causes the northern hemisphere to be hotter. During winter in the northern hemisphere, the tilt of Earth means the Sun's rays are on more of an angle when they reach Earth. This causes less heating, so the northern hemisphere is colder. The reverse happens for the southern hemisphere.

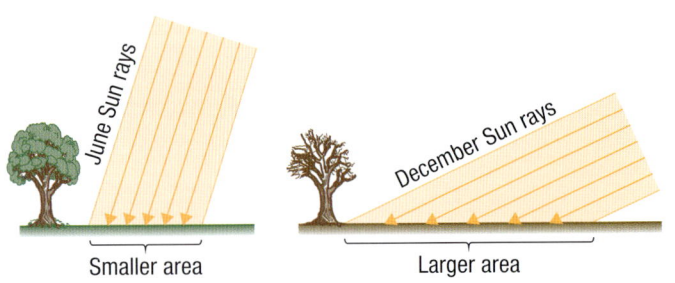

FIGURE 8.8 Why summer rays heat up Earth more than winter rays

FIGURE 8.9 (a) Summer and (b) winter views of the same road

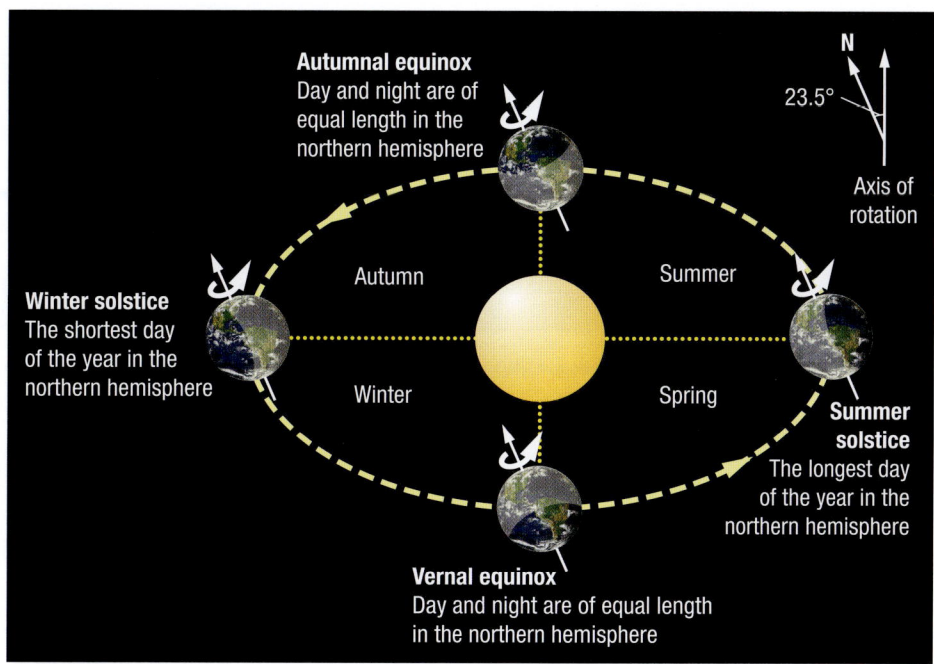

FIGURE 8.10 Explaining the seasons in the northern hemisphere

> **ACTIVITY** **Explaining the seasons to a student of your age**
>
> Write a full explanation of the seasons. It should be suitable for a student of your age. Emphasise the scientific principles involved using Figure 8.10 and the associated weblinks to help you.

Go to http://mypsci3.nelsonnet.com.au and click on **Seasons** to study the seasons in more detail.

Land of the midnight Sun

In the summer months of the northern hemisphere, Earth's tilt causes the Sun to never set in areas near the North Pole. At these latitudes (the Arctic Circle), there is all-day sunlight. In winter, the opposite happens and the Sun never rises – there is all-day darkness. The same happens near the South Pole at latitudes around the Antarctic Circle.

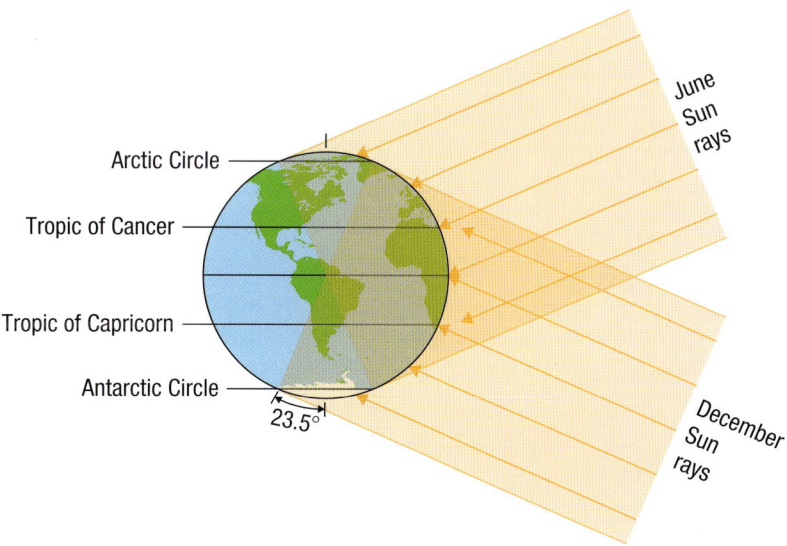

FIGURE 8.11 The Arctic and Antarctic Circles

FIGURE 8.12 Time-lapse photograph showing how the Sun never sets over a day inside the Arctic Circle

Effect of seasons on humans

Humans are not as affected by the seasons as other animals, but there are exceptions. Birth rates seem to be higher in summer and spring. Depression and seasonal affective disorder (SAD) are more common in winter in northern or southern countries with less sunlight. Conditions like SAD may be caused by disrupted melatonin levels, drops in the brain neurotransmitter serotonin, or disruptions to the biological clock. Doctors recommend the use of special lamps that mimic sunlight to help people overcome SAD.

How climate affects us

The latitude we live at clearly affects the climate we experience, the amount and intensity of sunlight, the lengths of days and nights, temperatures, rainfall and other weather patterns. The climate affects the way we live. It is part of our culture and heritage. It affects what we do in our free time, what we eat, the amount of energy we use, the work we do, when we work (for example, traditionally Spain has midday siestas), the design of cities (for example, the kilometres of walkways under the city of Toronto), our use of transportation and our health (respiratory diseases are more common in wet, cold climates, and malaria and yellow fever are more common in warmer climates). Different communities have adjusted to very different climates.

But this situation is changing. Cheap travel means more people can escape their home during winter to seek sunshine and warmth, thus experiencing two climates. More significant is global warming and the associated climate change, which will have significant impacts on many people's lives.

TA HELPING THE ELDERLY IN TIMES OF CLIMATE EXTREMES

Some elderly people struggle during times of extreme cold or extreme heat. You could investigate the degree of the problem in your area, the efforts local authorities are making to help people, the reasons for it and ways you could support people in this situation.

> **ACTIVITY** **Living in different climates**

Work in small groups to discuss the challenges associated with living in different climates. If possible, work in groups of students who come from, or have lived in, countries with different climates. Some stimulus questions include the following.
1. Which different countries and climates have you experienced?
2. How did you find going from your home climate to live in another climate?
3. How do you think climate affects people?
4. What are the advantages of hotter or cooler climates?
5. How does climate relate to cultural habits? Think of examples in your culture.
6. Do you think coming from different climates contributes to the cultural differences between people? Explain.

ATL

CRITICAL THINKING
The ability to synthesise and make connections between ideas to create new understandings

> **REVIEW**

1. Outline how Earth orbits the Sun.
2. Describe why the Earth is not always the same distance from the Sun.
3. State who Maui was in mythology.
4. Outline what was important about the idea of Sol Invictus.
5. Describe sunspots and their possible impact on Earth.
6. State what we mean by SAD. Outline how it can be cured.
7. Draw a labelled diagram to show how seasons are caused.
8. Describe the significance of the Arctic and Antarctic Circles.
9. Describe the purpose of the SDO project.

Influence of the Moon

> **ACTIVITY** **Basic information about the Moon**

Work in groups to produce a poster entitled '20 interesting facts about the Moon'. There are many suitable internet sites. You will find a suitable site on the accompanying weblink.

Go to http://mypsci3.nelsonnet.com.au and click on **Moon** to help you with research into the properties of the Moon.

The Moon is the biggest, brightest object in the night sky. It is Earth's only natural satellite. Poets and artists have been inspired by it and people have built temples to it. It looks different on different nights, changes shape and is always a source of mystery to us. The behaviour of some animals is affected by the phases of the Moon, and some people think human behaviour could be as well.

The origin of the Moon

A common theory for the origin of the Moon, called the giant impact hypothesis, is that it was created 4–5 billion years ago when a young, still-molten Earth was struck by another body about the size of Mars (Figure 8.13). There are also other theories, and variations on this theory. One of the most important pieces of evidence scientists have is rocks from the Moon. Analysis shows that the composition of the Moon's rocks is remarkably similar to those on Earth.

FIGURE 8.13 A theory to explain the formation of the Moon

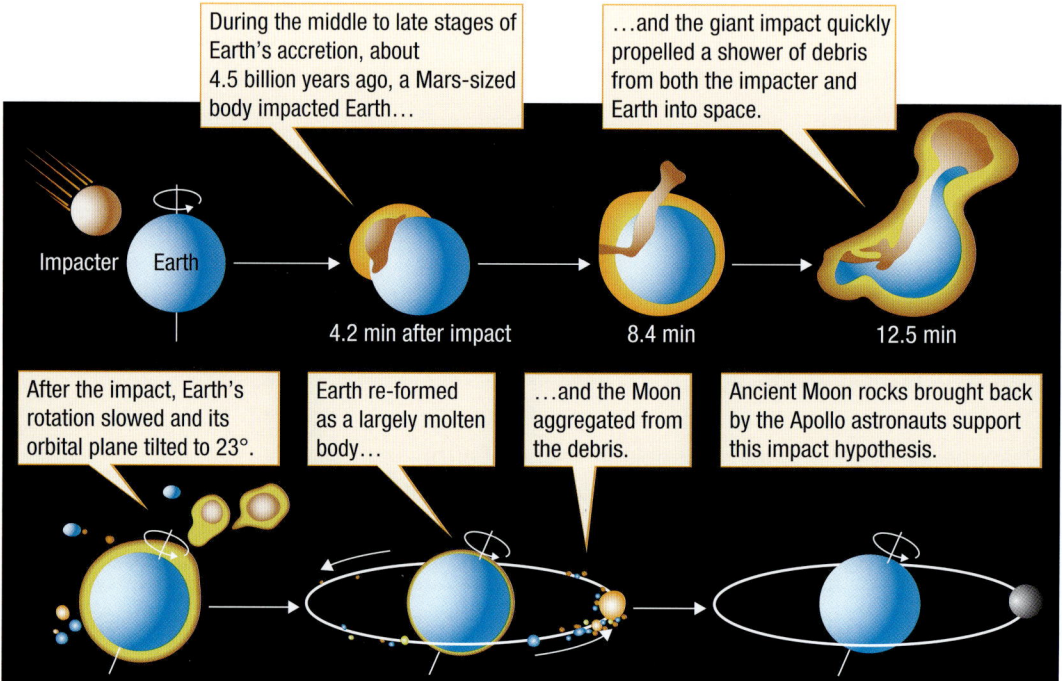

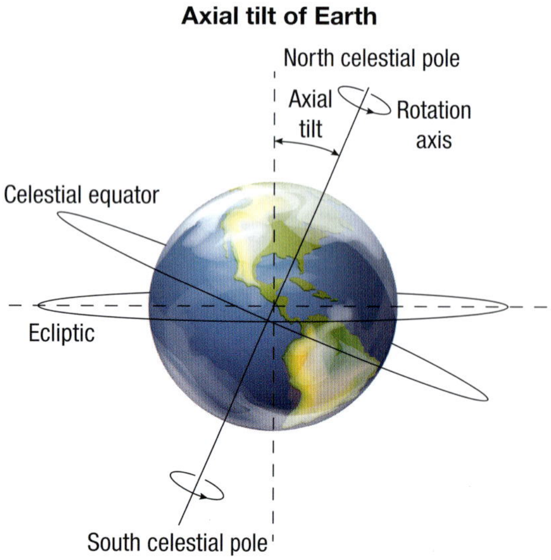

FIGURE 8.14 Earth's tilt

Effect of giant impact on Earth

It is thought that the collision with another large body gave Earth its present tilt (axis of rotation) of 23.5° (Figure 8.14). This tilt is the reason for different climates (see page 158). Also, it is thought that the collision slowed down the rotation of Earth so that we now have a 24-hour day, rather than perhaps a 5-hour day. Also, it is thought that the Moon's gravitational field has acted as a stabilising force on the rotation of Earth, so that the axis of rotation is always in the same direction. If not for the Moon, it is likely that the present North and South Poles would sometimes move to the Equator, and Earth would have suffered from even more extreme climate changes than it has.

Effect of the Moon on evolution

Although scientists don't fully understand the effect of the Moon on evolution, the existence of tides may have helped the development of life in water. The slowing of the rotation of Earth also probably favoured conditions for evolution of life as we know it. At least, it seems very likely that evolution would have been very different otherwise. Perhaps humans may not even have developed.

Effects of the Moon on Earth

The Moon's gravitational field clearly causes tides but it also has subtle effects on the movement of the oceans. Some scientists think it may be associated with phenomena such as El Niño ocean currents, which affect weather patterns around the world. It seems possible that the presence of the Moon may have had effects on Earth's molten core and could be involved in the movement of tectonic plates.

Effect of the Moon on living organisms

The Moon has influenced animal and plant life in many ways. Clearly, much of the life around shorelines is adapted to the ebb and flow of the tide. The eyesight of many animals, predators and prey is adapted to moonlight. Humans have traditionally organised their hunting and farming around the Moon, for instance the time for harvest. Traditionally, people would plan their walk across a city to make use of the available moonlight.

Scientists have shown that some animals are influenced by the Moon, although the effect is likely to be associated with the tides or simply more light. For example, it seems that some animals have a type of internal 'circalunar' clock that runs in synchrony with the Moon and tides. This applies to fiddler crabs, which become more active when the tide is out, as do marine iguanas in the Galapagos. African dung beetles roll their dung balls in straighter lines when the Moon is out. The eagle owl becomes more active as the Moon becomes fuller, but other owls become less active to avoid predators that might see them better. The reproductive cycles of some fish relate to the phases of the Moon, but again this is likely to be the result of tides. In the Great Barrier Reef every December, hundreds of species of coral spawn at the same time. It is thought that many factors cause this but that moonlight is the key factor that triggers the spawning. Some species of sea turtle wait for the full moon's high tide to ride waves onto shore so they can lay their eggs as far as possible up the beach.

FIGURE 8.15 Turtles coming ashore on a high tide to lay eggs are affected by the Moon.

People often comment that human behaviour is affected by the phases of the Moon – the word 'lunatic' comes from the Latin for Moon, *luna*. Teachers, nurses and police sometimes comment on how they notice more strange behaviour around the full moon. Some scientific studies claim to have shown some relationships, such as reduced sleep during a full moon, but overall most scientists doubt that our behaviour is affected by some mysterious force from the Moon.

Cultural impacts

Most ancient calendars were based on the **lunar** cycles, and many religious events relate to the lunar cycle, including Easter, Ramadan, Passover and Tet. Most earlier cultures had Moon goddesses, such as Coyolxauhqui, the Aztec Moon goddess.

In fiction, werewolves are men who transform into wolves on the full moon. This myth is part of many cultures.

FIGURE 8.16 The Aztec Moon goddess Coyolxauhqui being dismembered

FIGURE 8.17 Werewolves are fictional beings who are influenced by the Moon.

The Moon in religion and mythology

ACTIVITY

The Moon is culturally important for all people.

Many well-known mythologies include female or male lunar deities (gods). Research some of these lunar deities and consider why they were important. Present your findings to your class.

FIGURE 8.18 The Moon

FIGURE 8.19 The Temple of the Moon at Machu Picchu, Peru. The Moon is important in many ancient religions.

Moon exploration

The Moon is the only place in the universe, other than Earth, where human beings have walked. It is important to remember that before the 1950s, all of the information about the Moon was obtained from observing it with telescopes.

In 1969, the Apollo 11 mission landed on the Moon. Neil Armstrong and Buzz Aldrin spent a few hours taking photographs and collecting samples of **lunar** rocks. The amazing achievement of landing people on the Moon and bringing them back again fascinated the whole world and inspired many students to try to become astronauts. There were six manned landings from 1969 to 1972, and dozens of rovers and other craft landed. There are approximately 500 satellites presently orbiting the Moon. Some companies have advertised tourist journeys to the Moon.

FIGURE 8.20 Astronaut walking on the Moon

ACTIVITY: Impact on society of Moon landings

The images from the Apollo 11 Moon landing are likely to stay in society's consciousness forever. The purpose of the Moon landings was about more than putting a man on the Moon. Write a 300-word essay on what impact you feel the Apollo missions to the Moon have had on society.

Before you start, read about the Pale Blue Dot photograph and some of Carl Sagan's quotes about it. You will find many references to the Pale Blue Dot on the internet.

Go to http://mypsci3.nelsonnet.com.au and click on **Apollo, 40 years after** to read a newspaper article celebrating the Apollo program.

Phases of the Moon

Each night the Moon looks slightly different. On some nights it's a bright round disc, a full moon, and on other nights it's a thin crescent of light. The changes in its appearance aren't random. If you're lucky enough to have clear night skies for a few days (or even weeks), you will see how the Moon's shape appears to be slightly different each night. The Moon rotates on its own axis, just as Earth does. The time it takes to rotate once on its axis is the same as the time it takes to orbit Earth once. This means that from Earth we always see the same side of the Moon. Until the 1950s, when a Soviet spacecraft was sent to photograph it, humans had never seen the other side of the Moon.

From the tiny thin crescent (new moon), the Moon grows rounder and rounder until full moon. This is called **waxing**. Then the reverse happens as the Moon **wanes**, until one night it can't be seen at all. These changes in appearance are called the phases of the Moon. To understand these phases, it's important to remember two things: the Moon is not a luminous object and it doesn't actually change shape.

The phases are a result of the Moon's orbit around Earth and the way we see the sunlight that is reflected from its surface. In Figure 8.21, you can see the phases of the Moon.

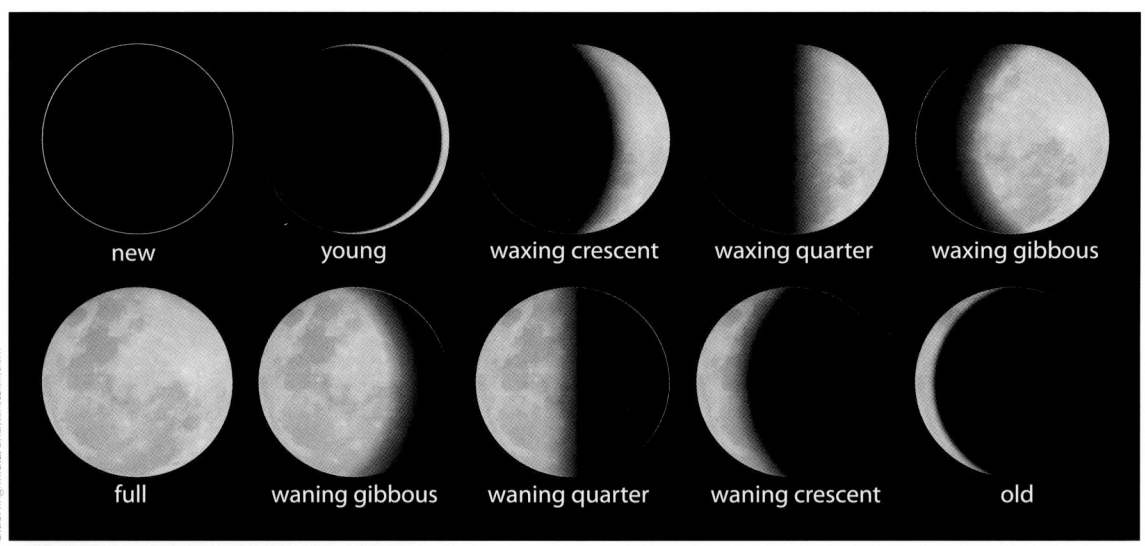

FIGURE 8.21 The phases of the Moon

FIGURE 8.22 A crescent moon

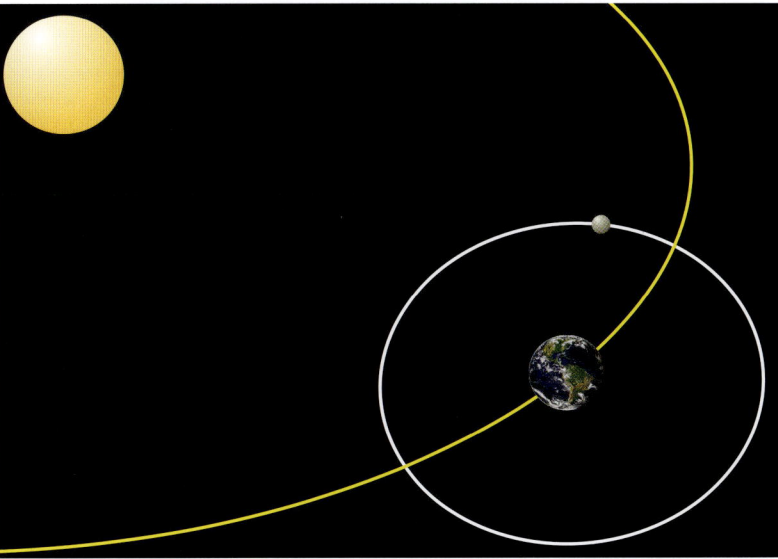

FIGURE 8.23 The relative positions and orbital paths of Earth, the Moon and the Sun (Earth and Moon to scale, but Sun not)

ACTIVITY Phases of the Moon

Use the associated weblink to help you complete this activity. Decide whether the following statements are true or false. Rewrite any false statements to make them true.

1. The average distance of the Moon's orbit around Earth is 3824 km.
2. The Moon is not a luminous object.
3. Apart from during eclipses, one half of the Moon is always lit up by sunlight.
4. Apart from during eclipses, one half of Earth is always lit up by sunlight.
5. The lit side of the Moon always faces towards Earth.
6. At a new moon, the unlit side of the Moon is facing Earth.
7. A **gibbous moon** is only seen when the Moon is waxing.
8. A full moon happens approximately 14 days after a new moon.
9. The Moon waxes for more days than it wanes.
10. The Moon's phases are due to Earth casting its shadow on the Moon.

Go to http://mypsci3.nelsonnet.com.au and click on **Phases of the Moon** to access and read further information about the Moon's phases.

REVIEW

1. Describe the orbit of the Moon around Earth. What holds the Moon in orbit?
2. Outline how scientists think the Moon was formed.
3. Describe how we think the formation of the Moon affected Earth.
4. Describe the influence the Moon may have on life on Earth.
5. State when the manned visits to the Moon took place.
6. Draw a labelled diagram to explain the phases of the Moon.

Eclipses

An **eclipse** takes place when the Moon or the Sun is partly covered (partial eclipse) or totally covered (total eclipse). The Moon orbits Earth. Earth orbits the Sun. At times during these orbits, all three line up exactly. When Earth is lined up between the Sun and the Moon, Earth's shadow falls across the Moon, causing a **lunar eclipse**.

FIGURE 8.24 A partial lunar eclipse

Occasionally, the Moon is exactly lined up between Earth and the Sun. When this happens the Moon blocks out the Sun and its shadow falls across Earth. This is a **solar eclipse**. A total solar eclipse is a truly dramatic event that is much rarer than a lunar eclipse.

Our ancestors (without any idea about the science behind the cause of eclipses) thought that eclipses signified that there was a disturbance to the natural order of things, that something was wrong. By 2300 BCE, astronomers were able to predict eclipses, although this knowledge was not widespread among ordinary people.

Solar eclipses can be either total or annular (when a ring of light is shown around the Moon), depending on the relative distance of the Moon from Earth. During a total eclipse, you can usually see flashes of light appearing on the outside of the Moon caused by light from the Sun passing between the mountains of the Moon. These are called Baily's beads. When just one Baily's bead is left, the eclipse looks like a shining diamond ring.

FIGURE 8.25 A total solar eclipse (one Baily's bead, the diamond-ring phase)

FIGURE 8.26 An annular solar eclipse

The Sun's diameter is about 400 times that of the Moon, but the Moon is about 400 times closer to Earth. This coincidence means that the Moon looks about the same size as the Sun and almost perfectly covers it during a total solar eclipse.

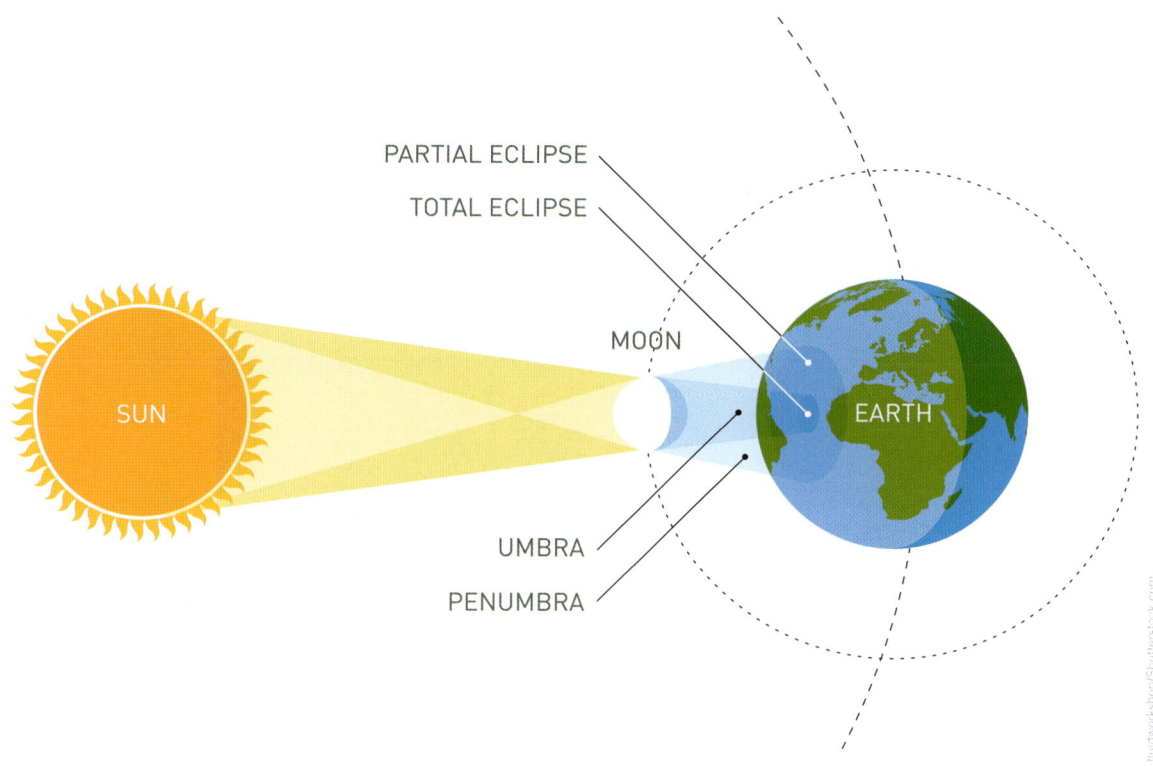

FIGURE 8.27 How solar eclipses take place

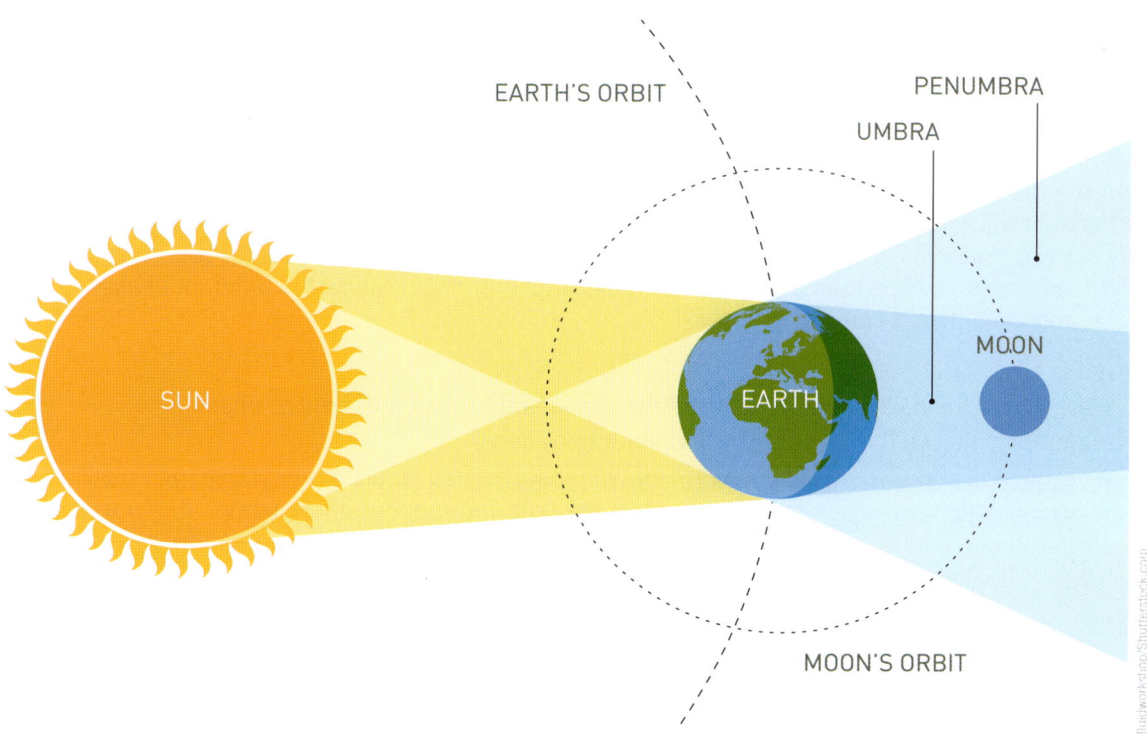

FIGURE 8.28 How lunar eclipses take place

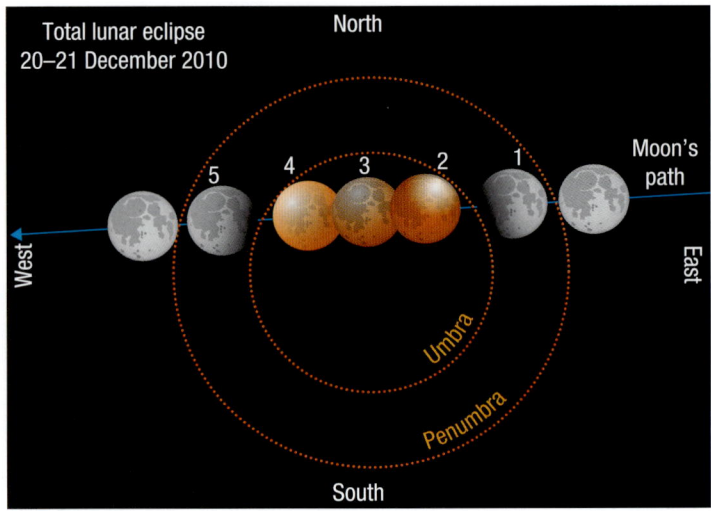

FIGURE 8.29 How the Moon changes as it goes from penumbra to umbra

ACTIVITY Modelling eclipses and phases of the Moon

YOU WILL NEED
- model-making materials
- light source

WHAT TO DO
In a group, design, set up and test a working model that shows how:
- solar eclipses occur
- lunar eclipses occur
- phases of the Moon occur.

You will present your model to the class.

THIS MIGHT HELP
Your teacher will arrange access to a darkened room or a black box and provide a suitable light source that will act as your 'Sun'. Various sized balls will also be available. Work out how you can model the movements of your 'Earth' and 'Moon' and then select suitable balls for your models. This might require some experimentation with the balls until they clearly show the different phases of the Moon and the different kinds of eclipses.

THINGS TO THINK ABOUT
- How big will the balls need to be to suit the space and to show the desired effects?
- How will you move the balls into their different positions?
- How will you make your presentation? Are you going to use labels or are you going to tell everyone what they are seeing as you move the balls around? Are there some other interesting facts you can tell the class during your presentation? Can you find some great images to liven up your presentation?

WHAT DO YOU THINK?
1. Describe two difficulties you had to overcome to get your working model right. How did you overcome them?
2. Of all the models constructed by the class, which seemed to show the phenomena of eclipses and phases of the Moon the most clearly? Which features worked well?
3. Scientists often use models to help explain phenomena they have observed. But models cannot perfectly show the real situation. We say that models have their limitations. One limitation to your working model was that the balls you used for Earth and the Moon were not to scale. Identify one other limitation.

Tides

The sea level changes during the day. When the sea comes in, it is called high **tide**. When it goes out, it is called low tide. The cause of the sea level change is the Moon. The **gravitational force** (gravity) of the Moon (and also, partly, of the Sun) pulls on Earth. You studied gravity in *Science 1 for the international student* Unit 6. This force that the Moon exerts on Earth causes the large bodies of water (the oceans) to bulge outwards. This bulge leaves behind another, slightly smaller bulge on the opposite side of the planet (Figure 8.30). Earth rotates on its axis every 24 hours, so all the parts of the planet pass through the two bulges, and there are two high tides each day.

Although the Moon is much smaller than the Sun, it is much closer to Earth and so its gravitational pull has a bigger effect on the tides than the Sun's gravitational pull. When the Moon and Sun are lined up to pull Earth's oceans in the same direction, it causes especially high tides. These are called spring tides (Figure 8.31). When the Moon and Sun pull at right angles to each other, the high tides are lower than normal. These are called neap tides (Figure 8.32).

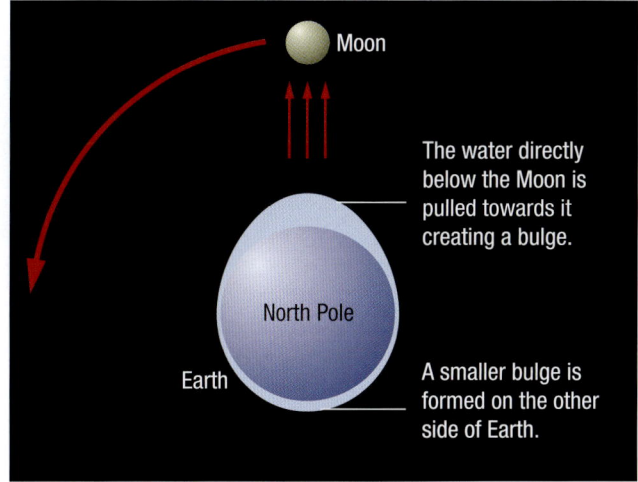

FIGURE 8.30 Explaining the tides

Out in the middle of the ocean, the effect of the tides is hardly noticeable. However, at some coastlines tides can cause enormous changes in sea level. The Bay of Fundy in Nova Scotia, Canada, has some of the biggest tides in the world. At times, the difference in sea level between high and low tide can be as much as 16 metres. That's about the same height as a five-storey building.

Anyone planning a sea journey needs to know about tide times. Tides do not just change the sea level, they can also result in strong currents that all sea goers such as fishers, sea-kayakers and swimmers need to be aware of.

The energy of all the moving water that causes the tides can be converted into electricity. This is called tidal power.

FIGURE 8.31 The alignment of Earth, Sun and Moon at spring tide

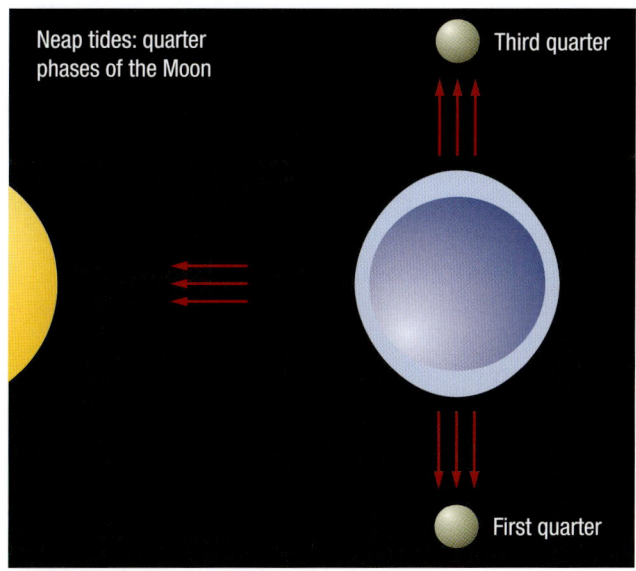

FIGURE 8.32 The alignment of Earth, Sun and Moon at neap tide

FIGURE 8.33 The same coastline at (a) high tide and (b) low tide

REVIEW

1. Draw two labelled diagrams to illustrate the difference between a solar eclipse and a lunar eclipse.
2. Outline what causes Baily's beads.
3. Which has the bigger influence on Earth's tides, the Sun or the Moon? Explain your answer.
4. Explain why there are two high tides every day.
5. Draw diagrams to show the relative positions of Earth, the Moon and the Sun at a:
    a neap tide
    b spring tide.
6. Suggest ways tides affect life on Earth.
7. Try to find out the origins of the names 'spring' and 'neap' for these special tides.

UNIT QUESTIONS

CRITERION A

EXPLAINING SCIENTIFIC KNOWLEDGE

1. Are the following statements true or false? Rewrite the false statements to make them true. (Level 1–2)
   a. The Moon is Earth's only natural satellite.
   b. The Moon's craters are the result of volcanoes erupting.
   c. The Moon is the same size as the Sun.
   d. There is no gravity on the Moon.
   e. The Moon has no atmosphere.
2. Explain what is meant by the 'equator'. (Level 1–2)
3. Organise the following statements to explain spring tides. (Level 3–4)
   I. This leads to exceptionally high tides called spring tides.
   II. When the Sun and Moon are lined up to both pull the oceans in the same direction, the water bulge is especially big.
   III. The bulge of water on one side also leaves a smaller bulge on the opposite side of the planet.
   IV. The gravitational pull of the Moon on Earth causes the water in the oceans on one side of the planet to bulge outwards.
   V. This means all areas experience two high tides every 24 hours.
   VI. The Sun's gravity also has an effect on our planet's water, but it isn't as big as the Moon's influence.
   VII. As Earth rotates, it passes through each bulge.
4. Outline how light from the Moon reaches Earth. (Level 3–4)
5. Outline what happens during a solar eclipse, and the difference between a total eclipse and an annular eclipse. (Level 5–6)
6. Describe the reasons for the different kinds of tides. (Level 7–8)

APPLYING SCIENTIFIC KNOWLEDGE AND UNDERSTANDING TO SOLVE A PROBLEM

7. How would you justify to someone that ancient societies worshipped the Sun? (Level 1–2)
8. What advice would you give a satellite about sunspots? (Level 3–4)
9. Suggest an implication for Earth if it were shown that the Moon is slowly orbiting further from Earth. (Level 5–8)
10. Suggest some implications for Earth if the angle of tilt of Earth suddenly increased. (Level 5–8)

INTERPRETING INFORMATION

11. You saw a lamp in a shop with an advertisement saying 'Don't be sad, use natural light'. What would your reaction be? (Level 1–4)
12. Look at Figure 8.34. Discuss as fully as you can the information shown in this diagram. Include explanations for the temperature differences. (Level 1–8)

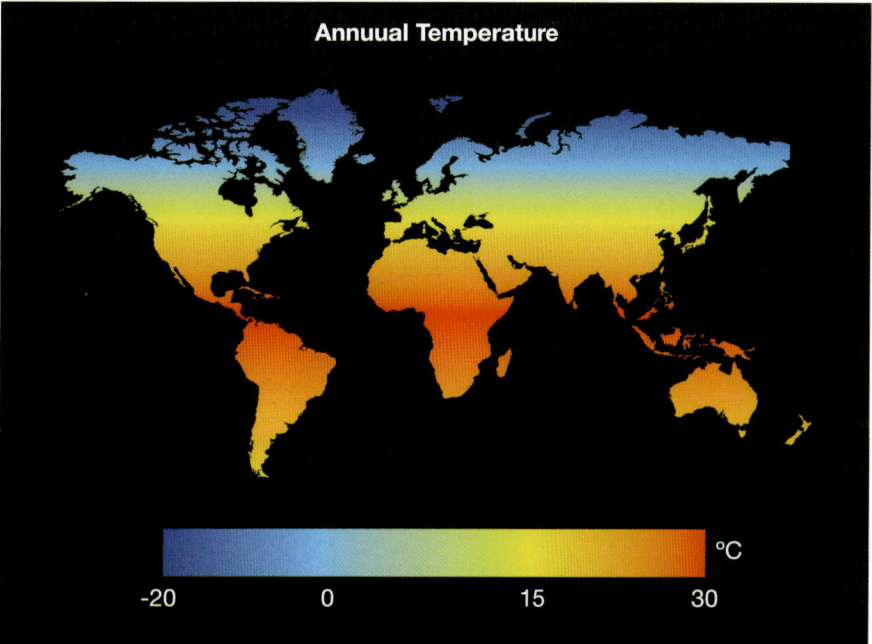

FIGURE 8.34 Annual mean temperature map

13 Look at Figure 8.35. Discuss as fully as you can the information shown in this tide calendar. Include explanations for the tides. (Level 1–8)

REFLECTION

1. In science, we often study the behaviour of systems. What does this mean? Why is a systems approach commonly used in science? Describe two different systems studied in this unit.
2. Explain why it can be useful to make a model to describe and explain a system.
3. Describe three different cycles of nature described in this unit. Name two other cycles of nature.
4. How can we use the term 'interaction' in relation to the Moon and Earth?
5. Do you believe we are influenced in some special way by the climate we live in?

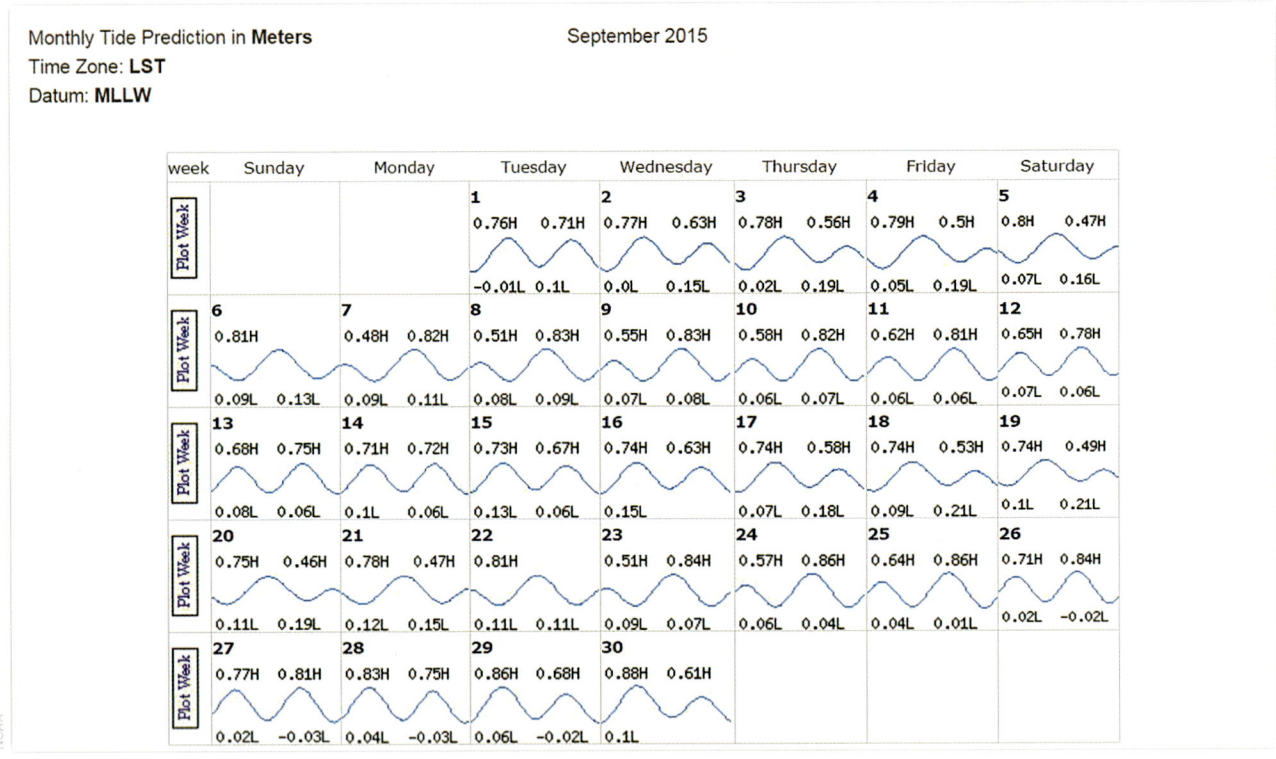

FIGURE 8.35 A tide calendar

Appendices

Appendix 1: Approaches to Learning (ATL) framework in MYP Sciences

One of the main reasons the IB is so respected worldwide is that it places Approaches to Learning (ATL) in a central role in all IB programmes, thus encouraging the skills, habits and dispositions necessary to succeed in learning, both at school and for the rest of your life. Whenever you are learning in the MYP, you should remember that there are two things happening.
1. You are learning about the subject you are studying.
2. You are learning about learning.

The key attribute you need is to believe that you can continue improving as a learner over your life. Throughout the *Science for the international student* books, you will see reminders of ATL skills (from the list below) that are involved in the tasks we have set.

ATL skill categories and clusters	ATL skills
COMMUNICATION I Communication	1 Express ideas clearly, precisely and persuasively. 2 Use effective and correct scientific language. 3 Use appropriately a variety of media for communication, including 21st-century technologies. 4 Use appropriate forms and modes of communication for different purposes and audiences. 5 Use strategies skilfully for speaking in public, reading for meaning, and structured writing.
SOCIAL II Collaboration	1 Show empathy to others when working in diverse teams, be aware of cultural differences, and encourage and support all members of a team. 2 Show flexibility and willingness to make necessary compromises to accomplish a common goal when working in groups. 3 Carry out a variety of roles and accept responsibility when working in groups/teams; show negotiating, advocacy, consensus-making and leadership skills. 4 Listen effectively and use non-verbal communication/body language. 5 Use social networking to build relationships effectively.
SELF-MANAGEMENT III Organisation	1 Understand the importance of setting personal goals, both long term and short term. 2 Manage time well, establish priorities and meet deadlines, using a daily and longer-term planner. 3 Prepare for and sit examinations, prepare a study program, make summaries, revise actively and control emotions. 4 Be organised for learning, including preparing materials, books, notes, online resources and necessary equipment for class. 5 Establish good support systems through family and friends, and create a pleasant place to study.
IV Affective	1 Be self-motivated, have a positive attitude, and believe you can improve as a learner. 2 Be mindful of mental distractions and how to improve focus. 3 Be resilient: cope well with failure and unexpected challenges. 4 Persevere in achieving long-term goals (have grit). 5 Be emotionally intelligent: control emotions and stress.

ATL skill categories and clusters	ATL skills
V Reflection	1 Be self-aware as a learner; be able to discuss your strengths and weaknesses and make goals for improvement. 2 Be able to give and respond well to feedback. 3 Show self-awareness of your learning, and be able to strategically plan how to carry out a task. 4 Monitor your work to review the progress being made, the areas of difficulty and the need for revised strategies. 5 Be knowledgeable about aspects of learning such as multiple intelligences and learning styles.
RESEARCH **VI** Information literacy	1 Access information from a range of sources in an efficient and effective way; be skilled in summarising information and note-taking. 2 Evaluate information critically; be able to identify primary and secondary sources, and identify points of view and bias. 3 Use information selectively, accurately and creatively for the task at hand. 4 Understand the legal and ethical implications around the use of information, academic honesty and intellectual property rights; use citations, footnotes, referencing and bibliographies. 5 Develop the skills to function in a knowledge economy, using digital technologies such as networking tools and social networks.
VII Media literacy	1 Think critically about media; analyse, evaluate and understand how and why media messages are constructed, and identify any bias, spin or misinformation that may be present. 2 Appreciate how individuals interpret messages differently. 3 Understand how media can influence our beliefs and behaviours, and culture and society generally. 4 Show an understanding of the ethical and legal issues surrounding the access to and use of media. 5 Understand and utilise the most appropriate media creation tools, characteristics and conventions to communicate information and ideas.
THINKING **VIII** Critical thinking	1 Use various types of reasoning, such as deduction and induction, as appropriate to the situation. 2 Be able to logically design scientific investigations to explore research questions, to develop an appropriate hypothesis and to control variables. 3 Reflectively analyse and evaluate evidence, data, arguments, alternative points of view, and claims and beliefs, to make valid judgments, conclusions, interpretations and decisions. 4 Synthesise and make connections between information and arguments to create new understandings. 5 Solve problems effectively, including in non-familiar situations; ask penetrating questions to clarify the problem.
IX Creative thinking	1 Show curiosity, a desire to dig deeper; enjoy novelty and uncertainty, and coming up with new ideas, products and solutions. 2 Be creative and imaginative, play with ideas; show divergent thinking, and be willing to let go and take risks; tolerate ambiguity; see mistakes as opportunities for learning. 3 Reason through metaphor and analogy; elaborate ideas; synthesise disparate bits of information; utilise knowledge in new contexts; formulate general concepts by abstracting common properties. 4 Design new products and technologies; be innovative and show entrepreneurial skills. 5 Create original works and ideas.
X Transfer (of skills and knowledge)	1 Show the motivation and meta-cognitive ability to support possible transfer of skills and knowledge within a discipline or across disciplines. 2 Be able to apply conceptual understandings and skills to new situations and across disciplines. 3 Appreciate the importance of interdisciplinary challenges and authentic problems, in which transfer of skills and knowledge is so important. 4 Be knowledgeable about recent developments in neuroscience and use information about the functioning of the brain to discuss learning (including the value of active, inquiry-based, contextual, collaborative and conceptual learning). 5 Understand how memory works and use techniques to improve memory.

Appendix 2: MYP Science 3 assessment criteria

	Achievement Level			
	1–2	**3–4**	**5–6**	**7–8**
Criterion A **Knowing and understanding**	i **Recall** scientific knowledge. ii **Apply** scientific knowledge and understanding to **suggest solutions** to problems set in **familiar situations**. iii **Apply** information to **make judgments**.	i **State** scientific knowledge. ii **Apply** scientific knowledge and understanding to **solve problems** set in **familiar situations**. iii **Apply** information to **make scientifically supported judgments**.	i **Outline** scientific knowledge. ii **Apply** scientific knowledge and understanding to **solve problems** set in **familiar situations** and **suggest solutions** to problems set in **unfamiliar situations**. iii **Interpret** information to **make scientifically supported judgments**.	i **Describe** scientific knowledge. ii **Apply** scientific knowledge and understanding to **solve problems** set in **familiar situations and unfamiliar situations**. iii **Analyse** information to **make scientifically supported judgments**.
Criterion B **Inquiring and designing**	i **State** a problem or question to be tested by a scientific investigation, with **limited success**. ii **Select** a testable hypothesis. iii **State** the variables. iv **Design** a method with **limited success**.	i **State** a problem or question to be tested by a scientific investigation. ii **Outline** a testable hypothesis **using scientific reasoning**. iii **Outline** how to manipulate the variables, and **state** how **relevant data** will be collected. iv **Design** a **safe method** and **select materials and equipment**.	i **Outline** a problem or question to be tested by a scientific investigation. ii **Outline and explain** a testable hypothesis **using scientific reasoning**. iii **Outline** how to manipulate the variables, and **outline** how **sufficient, relevant data** will be collected. iv **Design** a **complete and safe method** and **select appropriate materials and equipment**.	i **Describe** a problem or question to be tested by a scientific investigation. ii **Outline and explain** a testable hypothesis using **correct scientific reasoning**. iii **Describe** how to manipulate the variables, and **describe** how **sufficient, relevant data** will be collected. iv **Design** a **logical, complete and safe method** and **select appropriate materials and equipment**.

	Achievement Level			
	1–2	**3–4**	**5–6**	**7–8**
Criterion C **Processing and evaluating**	i **Collect and present** data in numerical and/or visual forms. ii **Accurately interpret** data. iii **State** the validity of a hypothesis **with limited reference** to a scientific investigation. iv **State** the validity of the method **with limited reference** to a scientific investigation. v **State limited** improvements or extensions to the method.	i **Correctly collect and present** data in numerical and/or visual forms ii **Accurately interpret** data and **describe** results. iii **State** the validity of a hypothesis based on the outcome of a scientific investigation. iv **State** the validity of the method based on the outcome of a scientific investigation. v **State** improvements or extensions to the method that would benefit the scientific investigation.	i **Correctly collect, organise and present** data in numerical and/or visual forms. ii **Accurately interpret** data and **describe** results **using scientific reasoning**. iii **Outline** the validity of a hypothesis based on the outcome of a scientific investigation. iv **Outline** the validity of the method based on the outcome of a scientific investigation. v **Outline** improvements or extensions to the method that would benefit the scientific investigation.	i **Correctly collect, organise, transform and present** data in numerical and/or visual forms. ii **Accurately interpret data** and **describe** results **using correct scientific reasoning**. iii **Discuss** the validity of a hypothesis based on the outcome of a scientific investigation. iv **Discuss** the validity of the method based on the outcome of a scientific investigation. v **Describe** improvements or extensions to the method that would benefit the scientific investigation.
Criterion D **Reflecting on the impacts of science**	i **State** the ways in which science is used to address a specific problem or issue. ii **State** the implications of the use of science to solve a specific problem or issue, interacting with a factor. iii **Apply** scientific language to communicate understanding **with limited success**. iv Document sources **with limited success**.	i **Outline** the ways in which science is used to address a specific problem or issue. ii **Outline** the implications of using science to solve a specific problem or issue, interacting with a factor. iii **Sometimes apply** scientific language to communicate understanding. iv **Sometimes** document sources **correctly**.	i **Summarise** the ways in which science is applied and used to address a specific problem or issue. ii **Describe** the implications of using science and its application to solve a specific problem or issue, interacting with a factor. iii **Usually apply** scientific language to communicate understanding **clearly and precisely**. iv **Usually** document sources **correctly**.	i **Describe** the way in which science is applied and used to address a specific problem or issue. ii **Discuss and analyse** the implications of using science and its application to solve a specific problem or issue, interacting with a factor. iii **Consistently apply** scientific language to communicate understanding **clearly and precisely**. iv Document sources **completely**.

*Factors include moral, ethical, social, economic, political, cultural or environmental.

Appendix 3: Guidance on carrying out and writing up MYP 3 scientific investigations (criteria B and C)

Title
- Give the investigation a title; for example, 'Properties of Springs'.

Problem/Research question (criterion B i)
- Outline what you are trying to find out; for example, 'How does the weight on a spring affect its length? I am interested in the type of relationship and whether the amount of weight used makes any difference to the length of the spring.'

Hypothesis (criterion B ii)
- Outline what you think is going happen in your investigation; for example, 'My hypothesis is that the spring will stretch as I put more weights on it.'
- Use your scientific knowledge to explain why you think this will happen; for example, 'I think it will stretch because a force will cause the particles of the spring to separate.'

Variables (criterion B iii)
- To do a fair test, you must only change one variable at a time.
- Write down the variable you will measure (dependent variable); for example, length of spring.
- Write down the variables that you will change (independent variables); for example, weights put on the spring.
- Write down the variables that you will keep the same while doing your investigation (control variables); for example, the spring you use.

Experimental method (criterion B iii and iv)
- List all the materials and equipment you will use.
- Design a logical, complete and safe method, showing the materials and equipment you have selected. Include a special section on the safety issues. Drawing a diagram will often help you explain what you want to do.
- Include what range of measurements you will take and/or the size of the sample, and how often you will repeat readings.
- Your plan also needs to include an explanation of how you will control and manipulate the variables and how you will collect sufficient and relevant data. Describe how you will process the data.
- Sometimes you might need to carry out a preliminary experiment to check that your plan actually works. You might need to alter some aspects of your initial plan.

Results (criterion C i)
- Collect, organise and present all your observations or measurements (data) carefully and fully; use a well-labelled table, including correct column headings and units. Sometimes you will be expected to transform your data to produce a new value, such as an average of results, or the speed of an object.
- Draw an appropriate graph of your results if they include numbers.

Conclusion and explanation (criterion C ii and iii)

- Accurately interpret your data; that is, what does your data tell you about your original research question? Comment on any patterns in your results. If you have numbers and a graph, discuss the shape of the line and the relationship it shows between the variables. This is your conclusion. Make sure you explain your conclusion, showing good scientific understanding.
- When writing your conclusion, construct a well thought out and reflective argument based on careful consideration of your evidence. Is your evidence good enough? If you feel your results don't really provide enough evidence to make a firm conclusion, then say so.
- Evaluate your prediction (hypothesis) – did it prove to be valid based on your results?
- Note: Where a result seems to be out of place and does not keep to the pattern, it is called an anomaly. (Perhaps a mistake was made in the reading.) You should discuss these results, and suggest reasons for their presence, but don't use them in making your conclusion.

Evaluation (criterion C iv and v)

- How well do you think your method worked? Did you have any problems you dealt with while carrying out the experiment?
- Validity: Did you collect sufficient valid data to answer the question? Did the instruments measure what they were meant to? What errors were there in your measurements? Was it truly a fair test? Did you repeat your readings enough times? Was your sample well chosen and large enough?
- Improvements: Write down how you might make your investigation better, especially to improve the validity or to obtain more reliable evidence. Write down any further experiments you could carry out to get more evidence or to extend this investigation.

Appendix 4: Articulating the conceptual framework in MYP Sciences

The MYP Science conceptual framework is defined by the key concepts of change, systems and relationships. These key concepts are further articulated through the use of related concepts. This conceptual framework will then be articulated in the curriculum via a series of integrating conceptual statements.

> In MYP Science, students develop their understandings about scientific **systems** and the **changes** that take place within these systems via the investigation of causal **relationships**.

The conceptual framework

Systems

Scientists use a systems approach to study the world. There are different kinds of systems, including the universe itself, the Earth, ecosystems, and closed systems in physics.

Related concepts	Integrating conceptual statements	Examples
Environment (biology)	The form, development and survival of an organism or a community is influenced by its surrounding **environment**, i.e. a combination of both abiotic and biotic factors.	In biology, we study the reasons for the changing populations of animals. In chemistry, we consider how pollution has affected fish life in lakes. In physics, we consider the impact of energy policies on the environment.
Interaction	Often there are **interactions** in science when two objects or more come together in a way that affects both of them.	In biology, we study the interaction between flowers and bees. In chemistry, we consider the interactions between water molecules. In physics, we study the interactions between electrostatic charges.
Models	Scientists use **models** to help them understand and to study some aspects of the world.	In biology, we use a model of the lungs to understand how breathing happens. In chemistry, we use models of chemical structures to understand the properties of salt (sodium chloride). In physics, we could use a model to simulate how particles move in a gas.
Cycles	To understand many systems, we need to consider **cycles** of energy and matter.	In biology, we consider cycles such as menstruation. In chemistry, we study cycles of elements in nature, such as carbon and nitrogen. In physics, we study cycles in the appearance of sunspots.
Scales	Different systems work at different **scales**, usually in relation to size, energy and speed. Scientists have developed the SI system of units to aid communication.	In chemistry and biology, we learn that different scales in sizes of particles of matter have an enormous effect on their properties. In physics, we consider speeds that range from a few centimetres per year (moving tectonic plates) to millions of metres per second (the speed of light).

Change

In science, change is viewed as the difference in a system's state when observed at different times. This change could be qualitative (such as differences in structure, behaviour or level) or quantitative (such as a numerical variable or a rate).

Related concepts	Integrating conceptual statements	Examples
Energy	**Energy** can be transformed from one form to another. Energy flows can be tracked through a system. The total amount of energy is conserved in closed systems.	In biology, we track energy flows through an ecosystem. In chemistry, we study how we can relate the heat given out in reactions to bond energies. In physics, we study how solar energy can be converted to electrical energy.
Transformation	Molecules, organisms, materials and energy can be **transformed** from one form to another.	In biology, we learn about genetic transformations in cells or the transformation as caterpillars change into butterflies. In chemistry, we learn about sand being transformed into glass. In physics, we study how energy can be transformed from one form to another.
Movement	Change and **movement** are at the heart of all natural systems, from the universe itself to the smallest cell.	The study of biology and chemistry depends on understanding the movement of particles. In physics, Newton's laws are used to explain the movement of objects.
Balance	Achieving a **balanced** state or equilibrium is an important idea in many sciences. Feedback loops are an important aspect of understanding many stable systems.	In biology, we consider how the body regulates its temperature. In chemistry, we learn about how reversible chemical reactions reach equilibrium. In physics, we consider how the forces on an object can be balanced.
Evolution	The natural world can be understood by considering **evolutionary** changes, some gradual, others sporadic.	We study the theory of biological evolution, which examines the changes in all forms of life over generations. We also can consider how technology evolves over time.
Conditions	Physical **conditions** influence chemical reactions and physical properties of materials.	In chemistry, we learn about the influence of factors such as temperature on the speed of reactions and how erosion of rocks is affected by changes in physical conditions.

Relationships

Relationships in science indicate the connections found among variables through observation or experimentation. These relationships can also be tested through experimentation. Scientists often search for the connections between form and function. Modelling is also used to represent relationships where factors such as scale, volume of data, or time make other methods impractical.

Related concepts	Integrating conceptual statements	Examples
Patterns	Recognising and seeking explanations for **patterns** in nature or collected data is the first step in scientific inquiry.	In biology, we consider the patterns in fossil records. In chemistry, we study patterns in the reactions of elements. In physics, we consider the patterns we see in the movement of the planets.
Cause and effect	The search for underlying causes of scientific phenomena is based upon establishing **cause and effect** relationships. We must always remember 'correlation does not necessarily imply causation'.	In biology, we could look for cause-and-effect relationships to help us learn how to reduce certain diseases. In chemistry, we consider the causes of increased erosion. In physics, we look for cause-and-effect relationships between force and changes in motion.
Evidence	Scientists use observations and data to develop **evidence** to support their claims, conclusions or answers to research questions. Scientists need to make careful arguments to justify their evidence.	In scientific investigations, we need to carefully consider the evidence when making conclusions; for example, did the results show that a heavier weight on the pendulum made it swing faster or slower or did it make no change?
Consequences	Making changes to systems (especially ecosystems) can have significant **consequences**; sometimes these can be predicted, other times they can be unexpected.	We consider the consequences of global warming, the possible consequences of nanotechnology, and the consequences of increased air traffic.
Form and function	The **function** (purpose, role, way of behaving) of a system or a structure is related to its **form** (the shape, relationships between the parts, composition and properties).	In biology, this is a crucial concept, with applications ranging from studying the arrangement of legs on an animal, to the shape of seeds, to the form and function of different cells. In chemistry, we consider how a substance's structure at the ionic or molecular level results in its actual properties. In physics, we discuss how to alter the design of a bike to improve its performance.
Development	Scientific understanding is being continuously **developed** via a continuous process of scientific investigation.	In biology, we consider how ideas about the cause and spread of diseases developed, and also the theory of biological evolution. In physics and chemistry, we study how ideas about the atom developed, especially in the late 19th and early 20th centuries.

Related concepts	Integrating conceptual statements	Examples
Creativity	The scientific endeavour is associated with a high level of **creativity**.	We experience the importance of creativity by studying the contributions of many truly creative scientists, such as Rutherford, Darwin and Mendeleev, through designing and carrying out our own experiments, and in the way we use our ideas in science to develop higher-level understandings.

Note: The following diagram has limitations. There are many other relations between key and related concepts, and between different related concepts, not shown in this diagram. However, we feel it could help teachers and students start developing a mental map of how the conceptual framework for Sciences links together.

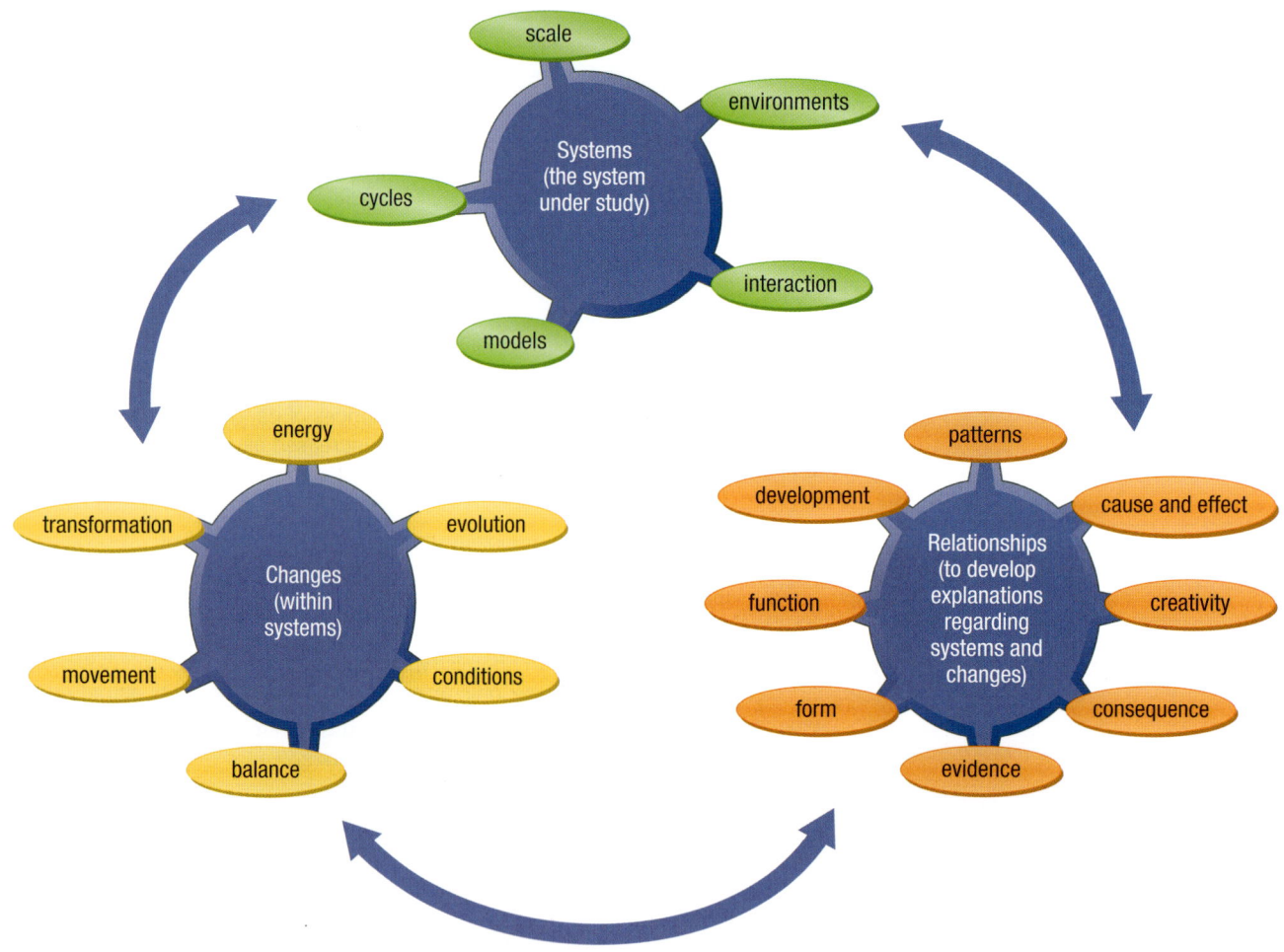

Glossary

acid a substance that tastes sour, reacts with bases and has a pH less than 7; in a reaction is able to give a hydrogen atom

acquired disability a disability that has arisen some time after birth, often as the result of an accident

adaptation a physical, functional or behavioural change to an organism that aids its survival

adolescence the period of human developmental growth that usually occurs during the second decade of life

alcoholism an alcohol over consumption disorder

alien species a species not native to the environment in which it is found

alkali metal a metal found in group 1 of the periodic table

alkaline earth metal a metal found in group 2 of the periodic table

alloy a substance with metallic properties that is a mixture of two or more elements

alveolus (plural alveoli) a tiny air sac in the lungs where gas exchange occurs

amplitude half the total height of a wave

ancestor an organism from which one is descended

antagonistic muscle pair two muscles that work in opposite directions

anorexia nervosa an eating disorder characterised by a distorted perception of body weight, abnormally low body weight and fear of gaining weight; potentially life threatening

artificial selection when humans select certain organisms as breeding partners in order to produce desired characteristics in offspring

assistive technology machines and control systems that make functioning easier, especially for people with a disability

asthma a chronic disease characterised by inflamed air passages, usually caused by airborne irritants

atom a tiny particle from which all substances are made; each element has its own kind of atom

balanced chemical or symbol equation an equation that uses chemical formulae to represent the reactants and products of a reaction and has the same number of atoms of each element on each side of the arrow

base a substance that tastes bitter, reacts with acids and, if soluble (alkali), has a pH greater than 7; in a chemical reaction is able to take a hydrogen atom

behavioural adaptation a change in an organism's behaviour that aids its survival

biceps a muscle in the upper arm

binge drinking the drinking a large amount of alcohol with the purpose of getting intoxicated

biopiracy when Indigenous knowledge is exploited by others without compensating the Indigenous peoples

bronchus (plural bronchi) a large delicate tube that carries air to smaller branches and cells in the lungs

bulimia nervosa an eating disorder where one may eat a large amount of food and then force vomiting; potentially life threatening

camouflage aspects of the colour or shape of an organism that help it hide in its natural environment

carbonate a compound containing carbonate, usually with a metal

carcinogenic causing cancer

cardiovascular system the organ system that transports blood around the body; consists of the heart and blood vessels

carnivore an organism that eats animals

chemical change (or chemical reaction) a change that takes place when chemical substances when chemical substances react with one another to produce new substances

chemical formula a shorthand way of showing the elements present in a substance and the relative numbers of their atoms

chemical reactivity the ability to react with other substances

chlorophyll the green pigment in plants' chloroplasts that is necessary for photosynthesis

chronic disease/illness a disease that is long-lasting or recurrent

cilia tiny hairs that move back and forth; found in the respiratory tract (nose, trachea, lungs), where they filter out foreign particles from air

climate the type of weather an area generally has over a long period of time

combustion another word for burning, or reacting with oxygen

combustion reaction a reaction in which a chemical reacts with oxygen and produces heat and light

community populations of organisms of different species that occupy a particular area, and usually interact with each other

companion planting the practice of planting certain plants next to, or close to, one another so that they can help each other grow better

compound a pure substance that is made up of two or more elements chemically combined

conductor of electricity a substance that allows electricity to pass from one end to the other
conductor of heat a substance that allows heat to pass from one end to the other
congenital disability a disability with which a person is born
corrosion the process of a metal reacting with the environment, especially oxygen, to produce a compound
cuticle a waxy layer on the surface of leaves

decomposition reaction a reaction in which a single reactant breaks down to produce more than one product
density the number of grams per cubic centimetre of a material
diaphragm a sheet of muscle between the chest and the abdomen that controls breathing
disability the incapacity to do what most people can do routinely
displacement reaction a reaction in which two elements swap or displace each other
dopamine a hormone released by the hypothalamus that is associated with the pleasure system of the brain; provides feelings of enjoyment and motivation to repeat the experience
driven gear the gear wheel that rotates when connected to a moving gear wheel (the driving gear)
driving gear the gear wheel that causes another gear wheel (the driven wheel) to rotate
ductile able to be drawn into wires

echo the reflection of sound from a solid surface
eclipse when the Moon (lunar) or Sun (solar) is partly or totally in shadow
ecology the science that examines the relationship between organisms and their environments

ecosystem all the living organisms and all the non-living parts, such as rocks, soil and water, that make up a particular environment
effort the force applied to a machine to achieve a result
electromagnetic spectrum the range of electromagnetic waves, which includes light
electroplating coating a metal by dipping it into a solution of the plating metal and passing an electric current through it
element a chemical made up of only one type of atom
ellipse a 'squashed' circle
endocrine system the collection of glands in your body that produce hormones
energy transfer the amount of energy needed to do work; energy transfer = force × distance
equilibrium position rest position, the place where a particle would be if it weren't disturbed by the energy of a wave
evolution changes in a species over a number of generations due to changes in DNA
extinction the complete loss of a species

follicle-stimulating hormone (FSH) a hormone produced by the pituitary gland that causes physical changes during puberty
fossil remains or traces of organisms that lived in the remote past
frequency (of a wave) the number of vibrations per second
fulcrum the point around which a lever turns; also known as the pivot
function what a body part or system does
functional adaptation a change within an organism's body that will aid the survival of that organism in its environment

gear ratio for two connected gear wheels, the ratio of the number of teeth on each wheel
gear wheel a wheel with teeth that fit into another toothed wheel
gibbous moon the phase of the Moon between the quarter moon and full moon phases
gravitational force the force that all objects exert on each other due to their mass; often just called 'gravity'

hardness resistance to being scratched
herbivore an organism that eats only vegetation
hertz (Hz) the unit used for measuring frequency; 1 Hz is one vibration per second
high quadriplegic a person with no muscle control below the neck
holistic looking at a system as a whole rather than simply considering the individual parts
hormones chemicals produced by the endocrine system that regulate many body functions, such as sleep, growth, mood and sexual function
hydrogen carbonate a compound containing hydrogen carbonate, usually with a metal

inclined plane a ramp
indicator a substance that changes colour with pH
Indigenous peoples peoples who have lived in a region for a very long time, and who have preserved their distinctive local traditions; the term also normally implies that their lands are presently governed by others
intellectual rights the legal protection given to the owner or creator of an intellectual idea or practice
International Classification of Functioning (ICF) an international agreement about disability

introduced species an organism that is not native to a certain area

involuntary muscles muscles over which the body has no control

kaleidoscope a tube, usually containing mirrors and coloured beads or similar, that makes interesting patterns

kinetic energy the energy possessed by an object due to its movement

latitude an imaginary circle on Earth parallel to the equator

law of conservation of mass a law that states that mass does not change during a chemical reaction; that is, atoms are neither created nor destroyed, simply rearranged during the reaction

limbic system six areas of the brain controlling emotions and formation of memories

limewater a substance that turns milky white when carbon dioxide bubbles through it

limewater test the test used to detect carbon dioxide gas

load the force that is applied against the effort and has to be overcome by the use of a machine

longitude an imaginary circle on Earth parallel to the circle that runs through the north and south geographic poles

luminous producing its own light

lunar related to the Moon

lunar eclipse when the Earth's shadow falls across the Moon

lung the major organ of the respiratory system

lustre (metallic) the mirror-like shine that is unique to metals

machine a tool that makes it easier to get something done

malleable able to be hammered into sheets

mammal an organism with hair that produces milk to feed its young

marsupial a class of mammals that develop their young in pouches

medium (plural media) the substance or material that a wave travels through

melting point (MP) the temperature at which a solid starts to melt (liquefy)

menstrual cycle the monthly cycle of changes in the female body, starting with the preparation of an egg for fertilisation and ending with the shedding of the lining of the uterus (bleeding)

metal an element located on the left-hand side of the periodic table that will conduct electricity and heat, and is opaque, shiny, malleable and ductile

metalloid an element located between metals and non-metals on the periodic table that possesses properties of both metals and non-metals

mineral a chemical element needed by organisms in small amounts and found in foods

molecule two or more atoms chemically combined

moment the turning effect of a force: moment = force × shortest distance between the force and the pivot

muscle soft tissue that can contract

musculoskeletal system the bones, muscles, cartilage, tendons, ligaments, joints and other tissues that hold the body's tissues and organs in place and allow movement

mutation a change of gene structure in the DNA of an organism that can be passed on to future generations

native something that occurs naturally in an area, without human intervention

natural gas a gaseous fossil fuel consisting mainly of methane

natural selection the process whereby environmental conditions help a species with particular characteristics to survive and pass on those advantageous characteristics to the next generation

neuron a specialised nerve cell that sends and receives electrical and chemical messages

neutralise add a base to an acid in the right proportions, or vice versa, to result in a neutral pH of 7

niche the special position of a species or population in its ecosystem

nocturnal only active at night

non-luminous does not produce its own light

non-metal an element found on the right-hand side of the periodic table that does not possess the properties of a metal

normal the range within which most people can operate routinely

nuclear fusion a reaction that releases energy when two nuclei join together to form a new, heavier nucleus

nutrient a substance needed by plants or animals for growth

omnivore an organism that eats both plants and animals

opaque not able to be seen through

orbit the circular or elliptical path of planets and satellites

ore a rock that contains metal

oscilloscope a device that can convert a sound wave into a waveform on a screen

parasite an organism that lives on or in an organism of another species (the host) and depends on the host for its nutrients

patent the sole right given by a government to the creator of an invention to make, use or sell the invention for a certain period of time

peak (or crest) the top of a wave
peer pressure the influence from others of the same age to change your attitudes or behaviour
periscope a device used to see above or around obstacles
pH an indication of the acidity of a solution; the more acidic the solution, the lower the pH; most solutions have a pH between 1 and 14
photosynthesis a chemical process used by plants, in which they convert water, carbon dioxide and sunlight to chemical energy (sugar), releasing oxygen
physical (or structural) adaptation a change in the physical structure of an organism that aids the organism's survival
physical change a change in which no new substances are formed
pitch the way we perceive the frequency of sound
pituitary gland a gland located in the brain that secretes a range of hormones, including FSH
pivot the point around which a lever turns; also known as the fulcrum
placenta a physical adaptation that enables the foetus of a class of mammals (placental mammals) to develop inside their mother's womb
plane (flat) mirror a shiny, flat surface that reflects light to make a clear image
plasma the 'fourth' state of matter that exists at very high temperatures
pneumatic switch a switch activated by movement of air
pneumatophore a physical adaptation in the form of a root that extends above the ground to aerate the underground root system of mangroves
pop test a test used to identify hydrogen gas; a 'pop' is heard when a lit splint or taper is held near the hydrogen gas

population a group of organisms of one species living in an area
pouch a physical adaptation in the form of a 'pocket' on the abdomen of female marsupials, in which their young develop
precipitate a solid that forms when two solutions are mixed
prefrontal cortex one of the areas of the brain responsible for personality, impulse control, decision making and planning
prescription drug a drug that a doctor prescribes a patient for a particular illness or condition, meant to be taken only by that patient according to directions
prey an organism that is eaten by a predator
principle of moments when a lever system is balanced, clockwise moment = anticlockwise moment
prism a transparent geometric block
product the new substance produced in a chemical reaction
puberty the time when sexual maturity occurs in individuals
pubic describes the area at the lower part of the abdomen, at the front of the pelvis
pulley a grooved wheel
purge vomit or take laxatives in order to get rid of food

random having no pattern
reactant the chemical substance that changes in a chemical reaction
reactivity series of metals a list of metals in order from the most to the least reactive
reflect bounce off
reproductive rate the rate at which new offspring are produced, increasing the population
respiratory system the body system that supplies oxygen to cells and disposes of carbon dioxide waste
respiratory surface the surface membranes of alveoli

rest position equilibrium position, the place where a particle would be if it weren't disturbed by the energy of a wave

salt a compound made up of a metal and a non-metal
satellite any object that orbits a star or a planet
simple machine one of five types of machine: inclined plane, lever, wheel and axle, pulley, gear wheels
skeletal muscles voluntary muscles, usually attached to bones
skeleton the system of all the bones of a human or animal
solar eclipse when the Moon passes between the Earth and Sun, casting part of the Earth into shadow
species a group of similar organisms that have the ability to reproduce to yield viable, fertile offspring
specific heat capacity the energy needed to heat 1 kg of a substance by 1°C
splint a thin piece of wood
state a phase of matter: solid, liquid, gas or plasma
stomata (singular stoma) small pores on the surfaces of plants that control gas exchange
structure the arrangement of the parts of something, such as a body part or system
sunspot a cool area on the Sun's surface
sustainable meeting the needs of the present generation without compromising the ability of future generations to meet their needs
synapse the junction between neurons, through which electrical and chemical messages are transmitted
synthesis reaction a reaction in which two or more reactants combine to form a single product

taper a long, thin candle

tarnish a layer on the surface of a metal produced by reaction of the metal with the environment which limits further reaction

tendon tough fibrous tissue

tensile strength the ability to withstand a stretching force

tide the change in sea level due largely to the Moon's gravitational pull

total internal reflection the process by which light is reflected from the inner surfaces of a prism rather than transmitted

trachea the tube that connects the mouth and nose to the lungs; also known as the windpipe

transition the process of changing from one state or condition to another, such as going from inside to outside

transition metal a metal found with in groups 3–12 of the periodic table

transmit pass through

triceps a muscle in the upper arm

trough the bottom of a wave

universal indicator a mixture of different indicators that displays a range of colours when added to solutions of different pH

vibration repeated movement backwards and forwards, from side to side or up and down

vitamin an organic substance required in small amounts for normal growth and activity of living organisms

vocal cords long tissues in the throat that vibrate to help us make sounds

volume (of sound) the loudness of a sound

waning reducing in size

wavelength the distance from peak to peak of a wave

waxing increasing in size

wheel and axle a solid or semisolid circle (wheel) through which a straight rod (axle) is connected

word equation a word summary to show the reactants and products of a chemical reaction

work the amount of energy transferred by a force that acts on something over a distance; work = force × distance = energy transfer

xerophyte a desert plant, such as a cactus, that has adapted to require less water

Index

A

acid–base reactions 104–5
acid–carbonate reactions 106
acid rain 78, 106
acids 102–3, 104
    metals reacting with 78, 103
acne 46
acquired disability 136
activity-related restrictions (people with disabilities) 134
adaptations 26–8
    bears and marsupials 29–31
    and Darwin's theory of evolution by natural selection 26, 32–3
    introduced species 31
    plants 27, 28, 37, 38, 39
addiction to drugs 55, 58
adolescence 45
    body image 46–7
    and the brain 54–5, 57
    peer pressure 53, 54, 57
    puberty 45–6
    regular exercise 48–50
    social you 53
adolescent health 44
    diet 50–1
    drug misuse and abuse 55–63
    eating disorders 47
    emotional health 52
    global challenge 44–5
    physical health 48–51
air particles 112
alcohol 56–7
alcoholism 56, 57
alien species 31
alkali metals 69, 103, 104
alkaline earth metals 70, 103
alloys 73
aluminium 76, 77
aluminium hydroxide 98
alveoli 60
American black bears 29
amplitude 114, 115
analogue signals 129
animals
    adaptations 27, 28, 29–31
    artificial selection 36
    evolution by natural selection 32–5
    human impact on 40
    impact of the Moon on 161
    introduced species 31, 40
anorexia 47
antacid tablets 104
antagonistic muscle pairs 49
Antarctic Circle 157
Arctic Circle 157
artificial nesting boxes 41
artificial selection 36
Asian black bears 29
assistive technologies 132, 134–5, 145–9
ATL (Approaches to Learning) skills and skill clusters 173–4
atoms 93
autism 135

B

Baily's beads 166
Baka Pygmies, Cameroon 22–3
balanced diet 11–12, 50
balancing chemical equations 93–5
bases 102, 104
batteries 78–80
bears, adaptations 29–30
behavioural adaptations 28
beryllium 104
bicarbonates 102
biceps 49
bicycle gears 143–4
binary code 128
binge drinking 56–7
bionic eye 134–5
biopiracy 19
birds
    adaptations 27, 33, 35
    Galapagos finches 35
blind people, assistive technologies 133, 134–5, 148–9
body image 46–7
boomerangs 10
brain
    and adolescence 52, 54–5, 57
    structures 54
brain waves and learning 129
brass 73
breathing in and out 61
bronchi 60
bronze 73
brown bears 29
bulimia 47
buying food grown locally 13

C

caesium 69
calcium 70
calcium carbonate 106
calcium hydroxide 98
calcium sulfate 106
camels, evolution 33
camouflage 29
cane toads 31
cannabis 57–8
carbon 73
carbon dioxide 37, 59, 61, 91, 96, 97, 100, 106
carbon monoxide 60, 101
carbonates 102, 106
carbonic acid 96
carcinogens 60
cardiovascular system 59
carnivores 29
carnivorous marsupials 31, 40
chamomile 16
chemical changes 88–9
    modelling 93
chemical equations 93
    balancing 93–5
chemical formula 102
chemical properties (of matter) 75
chemical reactions 88, 89
    gases produced in 91
    with metals 103–4
    observing 89–90
    predicting 100
    representing 93–5

types of 95–101
types of reactants 101–2
chemical reactivity 68
chlorine 73
chlorophyll 37
chronic disease 48
chronic fatigue syndrome 134
cigarette smoke 58–61
cilia 60
climate, affect on humans 158–9
climate change 35, 158
cobalt 70
cocaine 55
cochlear implants 113, 114
coefficients 93
combustion reactions 100–1
common ancestor 30
communication technology 110, 118, 128–9, 154
companion planting 9
compounds 76
conclusion and explanation (experiments) 178
congenital disability 136
conservation, and Indigenous peoples 22–3
cooking with heated stones 13–14
copper 70
copper carbonate 96, 97–8
corrosion 76–7
creation stories (Indigenous cultures) 8–9
crest 114
cultural values and differing world views 6–7
curanto 14
cuticle 37

D

Darwin, Charles 32
theory of evolution by natural selection 26, 32–3
deaf people 113, 117
decomposition reactions 96–7
density (metals) 68
diaphragm 61
diet
for adolescents 50–1
and Indigenous knowledge 11–13

digital signals 128
dingoes 31
disabilities *see* people with disabilities
displacement reactions 98–9
DNA 30, 36
dopamine 59
driven gear 143
driving gear 143
drug misuse and abuse 55–63
drumming language 118
dry environments, plant adaptations to 37, 38
ductility 67

E

ear, detection of sound waves 113
Earth 152
effect of giant impact on 160
influence of the Moon on 159–62
influence of solar storms on 154
influence of the Sun on 152–9
land of the midnight Sun 157–8
seasons 156–7
Earth's tilt 156, 157, 160
eating disorders 47
eclipses 166–8
effort (force) 139, 142
electrical conductivity 67
electrical control systems 145–9
electricity
making from metals 78–80
making in hydrogen fuel cells 96
electromagnetic spectrum 125–7
electromagnetic waves 125, 126–7
uses 127
electroplating 81–3
elements 66
elliptical orbit 152
emotions 52
endocrine system 45
energy transfer 111, 137
entrepreneurs 149
environmental change 35
equilibrium position 114
equinoxes 157
etching 103
evaluation (experiments) 178
evolution

by natural selection, Darwin's observations 26, 32–3
effect of the Moon on 161
and environmental change 35
and mutations 34
evolutionary tree of life 34
exhalation 61
experimental method 177
extinctions 31, 40
eyes 120

F

fair trade 21–2
fish
catching and preserving 12
in the diet 11
fish wheels 12
flat mirrors 121, 122, 123, 124
follicle-stimulating hormone (FSH) 45
food, Indigenous peoples' sourcing of 9–10, 12–13
fossils 30, 33
frequency (waves) 115, 125
friction 138
fulcrum 138, 139
function restrictions (people with disabilities) 133
functional adaptations 28

G

Galapagos finches 35
gamma rays 126
gases produced in chemical reactions 91
gasoline 101
gathering and growing 9
gear ratio 143
gear wheels 143–4
genes 32, 34
genetic modification 36
giant impact hypothesis 160
giant panda 30
global navigation satellite systems (GNSS) 148
Global Positioning System (GPS) devices 148, 149
global warming 158

glossary 183–7
gold 68, 70, 73, 76
group 1 metals 69, 103, 104
group 2 metals 70, 103
groups 3–12 metals 70

H

hangi 13–14
hardness (metals) 68
health
    Indigenous peoples 17–18
    see also adolescent health
hearing 113
heat conductivity 67
heliographs 121, 122
helium 73, 74
herbal medicines 15–16
herbivores 29
heroin 55
hertz 115
high tides 168, 169
hoists 145
holistic knowledge 6
hormones 45, 52
hot stones, cooking with 13–14
human ear, structure 113
human musculoskeletal
  system 48–9
humans
    effect of climate on 158–9
    effect of the Moon on
      behaviour 162
    effect of seasons on 158
    impact on natural
      communities 40–1
hunter-gatherers 9
hunting techniques 10
hydrochloric acid 102, 103, 104
hydrogen 73
hydrogen carbonates 102
hydrogen fuel cells 96
hydrogen gas 78, 91, 95, 103, 104
hydrogen sulfide 77
hydroxides 78, 98, 104
hypothesis 177

I

illegal drugs 55
inclined planes 137–8
indicators 104, 106
Indigenous knowledge 2–4
    about medicines 15–16, 19
    about navigation 20
    and diet 11–13
    as holistic knowledge 6
    and modern scientific
      knowledge 5
Indigenous peoples 2
    advantages of 'eating local' 13
    cooking with heated
      stones 13–14
    customs 3
    fair trade 21–2
    in film 22
    health 17–18
    impact of modern conservation
      approaches on 22–3
    and intellectual rights 18–19
    oral transmission of knowledge 8–9
    sourcing of food 9–10, 12–13
    sustainable lifestyle 3, 13, 20
    threats to, Kayapo (Brazil) 20–1
infant mortality, Indigenous
    cultures 17–18
infrared waves 126
inhalation 61
inherited characteristics 33
intellectual rights, Indigenous
  peoples 18–19
International Campaign to Ban
  Landmines (ICBL) 136
International Classification of
  Functioning (ICF) 133
introduced species 31, 40
involuntary muscles 50
iron 70, 76, 77

K

kaleidoscopes 123
kalua 14
kangaroos 31

Kayapo (Brazil), Amazon lifestyle
  under threat 20–1
kinetic energy 12

L

land bridges 30–1
land of the midnight Sun 157–8
landmines 136
law of conservation of mass 93–4
leaf hairs 37
leaves
    adaptation to dry environments 37
    controlling water loss from 38
LeBlanc, Terry, high quadriplegic
  sailor 147–8
legumes 9
levers 138–41, 142
light 120
    reflection from a mirror 120–2
    speed of 118
light waves 110, 118, 120, 125
    and sight 120
limbic system 52
limestone 106
limewater test 91
lithium 69
load (force) 139, 141–2
low tide 168
luminous objects 120
lunar eclipses 166, 167, 168
lunar rocks 160, 163
lung cancers 61
lung capacity 62
lungs
    breathing in and out 61
    structure and function 59–60, 61
    tobacco smoke effect on 59–61
lustre 67

M

machines, simple 137–44
magnesium 70, 103
magnesium chloride 103
magnesium hydroxide 98

malleability 67
mammals 30
mangroves 39
marsupial carnivores 31
marsupials, adaptations 30–1
medicines, Indigenous knowledge 15–16, 19
medium 111, 112, 113, 115, 120
melting point (MP) 68
menstrual cycle 46
metal displacement reactions 99–100
metalloids 66, 73
metals 66, 102
    alloys 73
    electroplating 81–3
    making electricity from 78–80
    in our body 66
    in our daily lives 69
    on the periodic table 66–7, 69–70, 72
    properties 67
    reactions 76–8, 103–4
    reactivity 69, 70, 75–6
    reactivity series 76, 100
    types of 69–70, 72
    variation in properties 68
methane 100
microwaves 126, 128
minerals 11
mining for chocolate 71
mirrors and mirror images 120–4
mistletoes 39
mobile phones 128
modelling chemical changes 93
modern communication technology 110, 128–9
modern science
    and Indigenous knowledge 5
    origins 4
molecules 93
moments 139–40
Moon 152, 159
    cultural impacts 162–3
    eclipses 166, 167, 168
    effect on evolution 161
    effect on human behaviour 162
    effect on living organisms 161–2
    effects on Earth 161
    exploration 163
    influence of the 159–63
    origin 160
    phases of the 164–5, 168
    and tides 161, 168–70
morphine 55
Mother Earth, creation stories 8–9
muscle cells 50
muscle pairs 49
muscles 49–50
muscular dystrophy 134
musculoskeletal system 48–9
mutations 34
MYP Sciences
    assessment criteria 175–6
    ATL skills and skill clusters 173–4
    conceptual framework 179–82
mythology, Sun and Moon in 153, 162

N

native wildlife 40, 41
natural gas 101
natural selection, evolution by 26, 32–3
navigation, Indigenous knowledge 20
neap tides 169
neem tree 19
neon 73, 74
neurons 54
neutralisation reactions 104–5
nickel 70
noise 116
non-luminous objects 120
non-metals 66, 67, 73
northern hemisphere 156–7

O

omnivores 29
one-pulley system 142
opaqueness 67
opiates 55
optical fibres 129
ores 71
organisms 26
    adaptations 26–31
    genetic modification 36
    in species variation 32–3
oscilloscopes 114
oxides 76
oxygen 59, 60, 61, 76, 77
oxygen gas 91, 95, 100

P

parasites 39
participation restrictions (people with disabilities) 134
patents 19
peak 114
peer pressure 53, 54, 57
penumbra 167
people with disabilities 132–3
    activity-related restrictions 134
    assistive technologies 134–5, 145–9
    classifications 133
    disabilities from wars 136
    participation restrictions 134
    physical and movement disability 136
    structure and function restrictions 133
periodic table
    metals on the 66–7, 69–70, 72
    non-metals and metalloids 66, 67, 73
periscopes 123–4, 125
peristalsis 50
petrol 101
pharmaceutical companies 19
phases of the Moon 164–5, 168
photosynthesis 37
physical adaptations 27
physical changes 88
physical or movement disability 136
physical properties (of matter) 75
pinhole camera 154–5
    size of image 156
pitch (sound) 113, 115–16
pituitary gland 45
pivot 138, 139
placental mammals 32
plane (flat) mirrors 121, 122, 123, 124
plants
    adaptations 27, 28, 37, 38, 39
    artificial selection 36
    evolution by natural selection 32

pneumatic switches 146–7
pneumatophores 39
polar bears 29
pollution 101
Polynesians 20
pop test 91
potassium 69, 104
potassium hydroxide 104
potato battery 79
predators 31
prefrontal cortex 54
prescription drugs 55
principle of moments 140
prisms 124–5
products 93
puberty 45–6
pubic hair 45
pulleys 141–2
purging 47

Q

quolls 31
Quovis electric car 135

R

radio waves 126
ramps 137, 138
reactants 93, 102–3
reactivity of metals 69, 70, 75–6
reactivity series of metals 76, 100
recycling 80
reflection 120
regular exercise 48–50
religions, Sun and Moon in 153, 162
research questions 177
respiration 61
respiratory surfaces 60
respiratory system 59–60, 61
rest position 114
results (experiments) 177
robotic arms 141
rusting and rust prevention 76, 77

S

sailing, assistive technology use 146–8
salmon 11, 12

salts 78, 104
scientific investigations, guidance on carrying out and writing up 177–8
scientific method 4
screws 137
seasons 156–7
    effect on humans 158
seesaws 138–9, 140
semiconductors 73
shifting cultivation 20
sight 120
silicon 73
silver 70, 76, 77
silver nitrate 98
simple machines 137–44
skeletal muscles 49
skeleton 49
slope, measuring 138
sloth bears 29
smoking tobacco 58–9
    effect on the lungs 59–61
    experiment 62
    and health 61
smooth muscles 50
social groups 53
sodium 69
sodium chloride 98, 104
sodium hydrogen carbonate 97
sodium hydroxide 78, 104
solar deities 153
Solar Dynamics Observatory 154
solar eclipses 166, 167, 168
solar storms, influence on Earth 154
solstices 153, 157
sound
    as a form of long-distance communication 118
    pitch and frequency 113, 115–16
    properties 112
    speed of 118, 119
    volume 113, 114, 116
sound waves 110, 111, 112–19
    distance travelled 117
    ear detection of 113
southern hemisphere 156
species extinctions 40
specific heat capacities 13
spectacled bears 29
speech disabilities 135
speed of light 118

speed of sound 118, 119
spring tides 169
state of the substance 93
steel 70, 73
stimulants 55
stomata 37
structural adaptations 27, 35
structure and function restrictions (people with disabilities) 133
sulfur 73
sulfur dioxide 77
sulfuric acid 102, 106
Sun 152
    cultural impacts 153
    eclipses 166, 167, 168
    influence on Earth 152–9
    and land of the midnight Sun 157–8
    measuring the diameter of 154–5
    and seasons on Earth 156–7
sun bears 29
sunspots 154
sustainable lifestyle 3, 13, 20
symbol equations 93, 94
synapses 54
synthesis reactions 95–6

T

talking drums 118
tarnish 77
Tasmanian devils 31
tendons 49
tensile strength 68
theory of evolution by natural selection (Darwin) 26, 32–3
'three sisters' (crops) 9
thylacines 40
tides 161, 168–70
titanium 70
Tlingit (North America) 11
    balanced diet 11–12
    catching and preserving food 12–13
tobacco, smoking 58–62
tolerance to drugs 55
total internal reflection 124–5, 129
total solar eclipse 166
trachea 60

transition metals 70
    colours 70, 72
triceps 49
trough 114
two-pulley system 141, 142

U

ultraviolet waves 126
umbra 167
United Nations Declaration on the
    Rights of Indigenous Peoples 18–19
universal indicator 104

V

validity 178
variables 177
vibrations 111, 112

visible light 126
vision impairments 133, 148
    see also blind people
vitamins 11–12
vocal cords 111
volume (sound) 113, 114, 116

W

wars, disabilities from 136
water 95, 96
    metals reacting with 78, 103–4
water waves 111
wavelength 114
waves 110–11
    electromagnetic spectrum 125–7
    light waves 110
    sound waves 110, 111, 112–19
    what are they? 111

wheelchairs 132, 133, 134, 135
    ramps for 137, 138
wheels and axles 142
word equations 93
work 137, 138, 142

X

X-rays 126
xerophytic plants 37

Z

zoos 40

NOTES

NOTES

NOTES

NOTES

NOTES